Czechoslovak Academy of Sciences

Solid-Liquid Phase Equilibria

Czechoslovak Academy of Sciences

Scientific Editor
Ing. T. Boublík, CSc.

Scientific Adviser
Doc. Ing. Č. Černý, CSc.

Solid–Liquid Phase Equilibria

Jaroslav Nývlt

ELSEVIER SCIENTIFIC
PUBLISHING COMPANY

Amsterdam — Oxford — New York 1977

Published in co-edition with
ACADEMIA, Publishing House of the Czechoslovak Academy of Sciences, Prague
Distribution of this book is being handled by the following publishers
    for the U.S.A. and Canada
Elsevier/North-Holland, Inc.
52 Vanderbilt Avenue
New York, N.Y. 10017, U.S.A.
    for the East European Countries, China, Northern Korea, Cuba, Vietnam and Mongolia
Academia, Publishing House of the Czechoslovak Academy of Sciences, Prague
    for all remaining areas
Elsevier Scientific Publishing Company
335 Jan van Galenstraat
P.O. Box 211, Amsterdam, The Netherlands

Library of Congress Cataloging in Publication Data

Nývlt, Jaroslav.
    Solid-liquid phase equilibria.

    Bibliography: p. 248
    1. Phase rule and equilibrium. I. Title.
QD503.N96   541'.34   75-33865
ISBN 0-444-99850-0

With 87 Illustrations and 17 Tables

© J. Nývlt, Prague 1977

All rights reserved. No part of this publication may be reproduced, stored in a retrieval system, or transmitted in any form or by any means, electronic, mechanical, photocopying, recording, or otherwise, without the prior written permission of the publishers

Printed in Czechoslovakia

CONTENTS

List of symbols	7
Introduction	13
1. Conditions for phase equilibrium	17
1.1. Single-component systems	17
1.2. Multicomponent systems	19
1.3. The Gibbs phase rule	26
2. Methods of expressing the composition of a system	29
2.1. Solutions and melts	29
2.2. Concentration units	30
2.3. Methods of graphical representation	31
2.3.1. Single-component systems	32
2.3.2. Two-component systems	32
2.3.3. Three-component systems	33
2.3.4. Four-component system	44
2.3.5. Multicomponent systems	49
2.3.6. The straight line rule and the lever rule	48
2.3.7. Enthalpy diagrams	50
3. Types of systems and their phase diagrams	55
3.1. Single-component systems	56
3.2. Two-component systems	60
3.3. Three-component systems	72
3.4. Four-component systems	79
4. Measurement of phase equilibria in condensed systems	85
4.1. Analytical methods	85
4.2. Synthetic methods	92
4.2.1. Polythermal methods	92
4.2.2. Isothermal methods	98
4.3. Dynamic method	99
4.3.1. Cooling curve	99
4.3.2. The kinetics of the growth and dissolution of crystals	103
5. Correlation methods for calculation of phase equilibria	106
5.1. Methods based on geometrical concepts	106
5.1.1. Orthogonal projection	107
5.1.2. Central projection	111
5.1.3. A modified method of central projection	121
5.1.4. Clinogonial projection	132
5.1.5. The symmetrical correlation method	135

5.2.	Methods based on thermodynamic concepts	140
	5.2.1. The method based on the electrolyte solution theory	141
	5.2.2. The teory of conformal ionic solution	147
	5.2.3. The Zdanovskii method and modifications	150
	5.2.4. Expansion of relative activity coefficients	157
6.	Impurity inclusion in crystals	176
	6.1. Isomorphous inclusion	176
	6.2. Adsorption inclusion	178
	6.3. Formation of anomalous mixed crystals	178
	6.4. The mechanism of internal adsorption	179
	6.5. Mechanical inclusions	179
7.	Literature	180
8.	Tables of solubilities of various substances in water	181
9.	Tables of solubilities in ternary systems (J. Nývlt and J. Wurzelová)	205
	Subject index	247

List of Symbols

The symbols used in this book are defined here; since a large number of quantities were employed, many symbols are used in order to denote different quantities in different sections of the text. In order to avoid confusion, the significance of such symbols is always explained when they appear for the first time and the number of the section in which the symbol is used is given. The dimensions of quantities are given only when they contribute to the definition.

Symbol	Chapter/Section	Significance
A	5.2	constant
A_{ij}	5.2.4	interaction constant
a	1, 5.2, 6	activity
a_i	2, 5.2.4, 9	concentration (g per 100 g of solvent)
a_1, a_2	5.1.1	functions defined by eqns. 5.9 and 5.11
B	5.2.4	function defined by eqn. 5.196
B	5.2.1	constant
C_i	2, 5.2	molar concentration
c_i	2	concentration (g per 100 ml of solution)
c_1, c_2	5.1.1	function defined by eqns. 5.10 and 5.12
$c_1(y)$	5.1.1	integration constant
D	5.2.1	constant
D	6	homogeneous distribution coefficient
d		total differential
e	5.2.1	electron charge
F	5.1.3	function defined by eqn. 5.56
F	5.2.4	function defined by eqn. 5.197
f	1.2	fugacity
f	1.3, 2	number of phases
G	5.1.3	function defined by eqn. 5.57
G		Gibbs free energy
G		molar free energy
$\overline{G}$		partial molar free energy
g		weight
H		molar enthalpy
$\overline{H}$		partial molar enthalpy

Symbol	Chapter/Section	Significance
ΔH_{me}		molar heat of fusion
ΔH_{sol}		molar heat of solution
ΔH_{vap}		molar heat of vaporization
ΔH_∞		first differential heat of solution
i, j		summation symbols
K	5.2	association constant
k	5.2	Planck's constant
M		molecular weight
m_i	2, 5.2	molality
m	5.1.1	coefficient defined by eqn. 5.15
N_i	2	normality
N_A	5.2	Avogadro's constant
n	1	number of moles of a component
n	5.1.1	coefficient defined by eqn. 5.14
n	5.1.2	number of components
n	5.2	constant
P		pressure
P_o	1	solvent vapour pressure
p_i		concentration (wt.-%)
Q, Q_{ij}, Q_{ij}	5.2	constants
Q_{ij}	5.2.4, 9	interaction constants
q	1	heat
R		gas constant
r	5.2.1	distance from an ion
r^*	5.2.1	minimum distance between ions
r_o	5.2.1	radius of the ionic atmosphere
S	1	entropy
S	5.2	analytical solubility product
S_a	5.2	thermodynamic solubility product
S		molar entropy
$\overline{S}$		partial molar entropy
s		number of components
s_o	2.3.3	length of the side of a triangle
T		absolute temperature
T_{me}		absolute melting point
t		temperature (°C)
t_i	3, 5.1	temperature at which a solution of the given composition is just saturated with the i_{th} component
U	1	internal energy

Symbol	Chapter/Section	Significance
V	1, 2	volume
v		molar volume
$\bar{V}$		partial molar volume
V_{ij}, V_{ij}^*	5.2	salting-out volume
v	1, 2	number of degrees of freedom
W	5.2	work for the formation of the ionic atmosphere
w	1	work
w_i	2	weight fraction
X	5.2.3	concentration
x	1, 2, 5.2.3	molar fraction
x	5.1.1	coordinate
x	5.1.2—5.1.4	weight or molar fraction
x_i	5.2.4	relative molality defined by eqn. 5.171
x	6	concentration of micro-component
Y	5.2.3	concentration
y	5.1.1	coordinate
y	1, 2, 5.2.3	molar fraction
y	5.1.3	functions defined by eqns. 5.59 and 5.60
y	6	concentration of macro-component
y_i	5.2.4	molar fraction of the ith component in the solid phase
Z	5.2.3	concentration
Z	5.2.2	lattice coordination number
z	5.1.1	coordinate
z_i	2, 5.2	ionic valence
α	1, 3	symbol for a given modification
α	5.1.2	constant defined by eqn. 5.42
α	5.2.4	constant defined by eqn. 5.198
α_i	5.2.3	weight fraction of the ith solution in a mixture
β	1, 3	symbol for a given modification
β	5.1.2	constant defined by eqn. 5.43
β	5.2.4	constant defined by eqn. 5.199
γ		activity coefficient
Δ		change, difference
Δ	5.1.2	correction term
δ	1	slope of a straight line
$\varkappa$	1	function defined by eqn. 1.56
$\varkappa$	5.2.1	solvent dielectric constant
λ	5.2.2	energy parameter
λ	5.2.4	function defined by eqn. 5.166

Symbol	Chapter/Section	Significance
λ	6	logarithmic distribution coefficient
μ		chemical potential
μ°		standard chemical potential
γ	5.2	number of ions formed from one molecule
ξ_i	5.2.4	relative activity coefficient
π		Ludolf number
ϱ	2	density
$\sum$		sum
σ	5.1.1	symbol for solubility curve
σ_i	5.2	entropy term
τ		time
ψ	5.2.4	function defined by eqn. 5.186
$\varnothing$	5.2.4	function defined by eqn. 5.173 or 5.202
φ'	5.1.1	symbol for an unknown functional dependence
φ_i	2	volume fraction
ψ	5.2.4	function defined by eqn. 5.209
[ ]		composition of the system; concentration; constant conditions
$\overline{AB}$		segment, straight line passing through points A and B
∂		partial differentiation

Superscripts

Symbol	Chapter/Section	Significance
$''' (f)$	1, 2	symbols for phases
$*$	5	ideal behaviour
o		standard state
E	5.2	excess function
i, j, l, 2, ... s		symbols for various components
l	1	liquid phase
o	2	solvent
oi	5.1	pure ith component
oi	5.2	binary solution
$\alpha\beta$		modification change
s		solid phase
sl		fusion
sat		saturated solution
lg		vaporization
α, β		symbols for modifications
$+$		cation
$-$		anion

Acknowledgement

I thank the management of the Research Institute of Inorganic Chemistry, Ústí nad Labem, and especially my co-workers in the crystallization laboratory, who helped me during the collection and preparation of the extensive material necessary for writing this book. I am also grateful to the Institute of Chemical Process Fundamentals of the Czechoslovak Academy of Sciences for the support given to this work and to Professor E. Hála, Dr. T. Boublík and Dr. Č. Černý, for valuable comments. In addition, I thank Dr. F. Moudrý, L. Provazník, J. Wurzelová and M. Jánová for their help in assembling the tabulated data.

J. Nývlt

INTRODUCTION

In order to characterize systems that contain various phases in contact with each other, data on the appropriate phase equilibria are necessary. These data, on the one hand, yield basic information on the behaviour of the system and, on the other, constitute a basis for determining the material balance, thus permitting further characterization of the system and of the processes taking place.

On the basis of the types of phases in contact, systems can be classified as

(a) gas-liquid
(b) gas-solid
(c) liquid-liquid
(d) liquid-solid
(e) solid-solid

In the literature, most attention has been paid to group (a); for both gas solubilities and liquid-vapour equilibria, there are numerous papers and monographs that treat this topic in sufficient depth. There are extensive collections of data on individual systems, experimental methods have been developed in detail and correlation methods and procedures for checking the reliability of experimental data are available. Finally, procedures for assessing the data in these phase equilibria have been described and can be used when experimental data are not available.

Gas-solid equilibria involve virtually only sublimation equilibria (if adsorption equilibria and interactions that involve chemical reactions are ignored) and they can therefore be treated in a similar manner to that employed for liquid-vapour equilibria.

Liquid-liquid equilibria are important chiefly for extraction processes and have been discussed in the literature.

A more complicated situation exists for solid-liquid equilibria. A thermodynamic treatment has long been known and appropriate experimental methods have been developed. However, although there are extensive tables of equilibrium data and various methods for the correlation and assessment of the data on phase equilibria of this type, no publication is available that treats all aspects of the field systematically.

Of a number of publications that deal with various aspects of phase equilibria, several books that have become classics should be mentioned:

Finally, solid-solid equilibria can be treated by using methods similar to those employed in the above instances, taking into account the specific features of these systems (limited contact between the phases, low particle mobility and the consequent very slow equilibration).

The study of solid-liquid equilibria has been an inseparable part of the basic research on crystallisation processes carried out in the Research Institute of Inorganic Chemistry. Over the years, a large body of experimental data has been assembled and various methods for handling experimental data and for the assessment of phase equilibria in multicomponent systems on the basis simple data have been verified and developed. However, the data and methods are scattered in the literature and I therefore consider it useful to collect some of the more important information in a monograph, especially as a basic knowledge of phase equilibria is useful for the solution of various technological problems.

The basic problem in determining phase equilibria in multicomponent systems is the existence of a large number of variables, necessitating extensive experimental work. If ten measurements are considered satisfactory for acceptable characterization of the solubility in a two-component system in a particular temperature range, then the attainment of the same reliability with a three-component system requires as many as one hundred measurements; with more complicated systems, the necessary number of measurements will be several orders of magnitude larger. Therefore, a reliable correlation method that would permit a decrease in the number of measurements necessary on the basis of dependences among concentration data, or would even make possible the assessment of solubilities in multicomponent systems by using the data available for simpler systems, would be extremely useful.

Among such methods, a correlation method based on a thermodynamic description of phase equilibrium with the concentration dependence of the relative activity coefficients expanded in a McLaurin series has proved suitable. An advantage of this method is that the empirical coefficients (which do not depend too strongly on the temparature) readily enable concentration and temperature interpolations to be made and, with a certain risk, also small extrapolations; more important, certain combinations of these coefficients also permit the expression of phase equilibria in systems with a greater number of components.

This method has been developed in two modifications: the simpler is valid for

G. Tammann: *Lehrbuch der heterogenen Gleichgewichte*, Braunschweig, 1924.

W. C. Blasdale: *Equilibria in Saturated Salt Solutions*, Chemicals Catalog Co., New York, 1927.

F. F. Purdon and V. W. Slater: *Aqueous Solutions and the Phase Diagram*, Arnold, London, 1946.

A. Findlay and A. N. Campbell: *The Phase Rule and its Applications*, Longmans, London, 1938.

systems with components that are immiscible in the solid phase,[1] while the more general method holds for systems that are partially or completely miscible in the solid phase.[2]

In order to facilitate the application of the correlation method to multicomponent systems, we have compiled the literature[3,4] data on solubilities in three-component systems with components immiscible in the solid phase and, on this basis, we have assembled tables of the constants of the correlation equation.

The tables of constants are further supplemented with solubility tables in binary aqueous systems, which also contain data on phase transitions and heats of solution for individual compounds.

[1] J. Nývlt: Collection Czechoslov. Chem. Commun. *34*, 2348 (1969).
[2] J. Nývlt, A. Majrich, H. Kočová: Collection Czechoslov. Chem. Commun. *35*, 165 (1970).
[3] F. W. Linke (A. Seidell): *Solubilities of Inorganic Compounds*, Van Nostrand, New York, 1958; Am. Chem. Soc., Washington 1965.
[4] *International Critical Tables*, McGraw-Hill, New York, 1926.

1. CONDITIONS FOR PHASE EQUILIBRIUM

The application of basic thermodynamic postulates to a closed system under constant temperature and pressure results in the phase equilibrium condition

$$dG = 0 \quad [P, T] \quad (1.1)$$

where G is the free energy of the system. If the system consists of several phases (denoted by ′, ″, etc.), the change in the overall free energy of the system is the sum of the changes in the free energies of the individual phases

$$dG = dG' + dG'' + \ldots + dG^{(f)} = 0 \quad [T, P, n] \quad (1.2)$$

1.1. Single-Component systems

In a closed, single-component system, mass can be transported among individual phases. The temperature, T, the pressure, P, and the number of moles, n, in a given phase will then determine its state.

$$dG' = \left(\frac{\partial G'}{\partial T}\right)_{P,n} dT + \left(\frac{\partial G'}{\partial P}\right)_{T,n} dP + \left(\frac{\partial G'}{\partial n'}\right)_{T,P} dn' =$$
$$= -S' \, dT + V' \, dP + G' \, dn' \quad (1.3)$$

Similar equations hold for the other phases.

By the simultaneous solution of these equations, an intensive criterion for equilibrium in a single-component system is obtained

$$G' = G'' = \ldots = G^{(f)} \quad (1.4)$$

For two-phase equilibrium between solid and liquid phases the general equation, eqn. 1.4, can be simplified to give

$$G' = G'' \quad (1.5)$$

or

$$dG' = dG'', \quad (1.6)$$

The molar free energies are, of course, functions of temperature and pressure so that, analogous to eqn. 1.3,

$$dG' = -S' \, dt + V' \, dP \quad (1.7)$$

and

$$dG'' = -S'' \, dT + V'' \, dP \quad (1.8)$$

where S and V are the molar entropies and molar volumes of the substance in the appropriate phases, respectively.

The basic relationship for the phase equilibrium condition can be derived from eqns. 1.6—1.8:

$$-(S' - S'') \, dT + (V' - V'') \, dP = 0 \quad \text{[equil.]} \tag{1.9}$$

or, after rearrangement,

$$\frac{dP}{dT} = \frac{S'' - S'}{V'' - V'} = \frac{\Delta S}{\Delta V}. \tag{1.10}$$

For phase equilibrium, it simultaneously holds that

$$\Delta S = \frac{\Delta H}{T} \tag{1.11}$$

where ΔH is the enthalpy change during transfer of one mole of the substance from phase ' to phase ", and therefore the Clapeyron equation is obtained

$$\frac{dP}{dT} = \frac{\Delta H}{T \Delta V}. \quad \text{[equil.]} \tag{1.12}$$

If, in a single-component system, equilibrium between the liquid and solid phases is considered, i.e. the melting or solidification of the substance, the Clapeyron equation can be rewritten to give

$$\frac{dP}{dT_{sl}} = \frac{\Delta H_{sl}}{(V_l - V_s) \, T_{sl}} \tag{1.13}$$

where ΔH_{sl} is the molar heat of fusion, V_l and V_s are the molar volumes of the substance in the liquid and solid phases, respectively, and T_{sl} is its melting point.

The V_l and V_s values are usually very close so that (dT_{sl}/dP) is very small and the effect of pressure on the melting point is therefore also small. With condensed systems, the effect of pressure on the state of the system under normal conditions (i.e. at pressures close to atmospheric) can be neglected.

Note: The equation must be dimensionally homogeneous; if ΔV is expressed in millilitres, ΔH must have the dimension ml. Pa/mole where 1 J = 9.863 ml. Pa.

According to the Richards rule, the entropy of fusion, $\Delta S_{sl} = \Delta H_{sl}/T_{sl}$, is approximately constant for related compounds and usually has the following values:

9.2 for metallic elements;
20.9 — 29.3 for inorganic compounds;
37.7.— 58.6 (often 56.5) for organic compounds.

This rule enables the latent heat of fusion for similar substances to be assessed.

For phase equilibrium between two solid phases in a single-component system, i.e. an allotropic change, the Clapeyron equation can be rewritten in the form

$$\left(\frac{dP}{dT}\right)_{\alpha,\beta} = \frac{\Delta H_{\alpha,\beta}}{(V_{s\alpha} - V_{s\beta})\,T_{\alpha,\beta}} \tag{1.14}$$

where $\Delta H_{\alpha,\beta}$ is the latent heat of the change (the difference in the heats of sublimation of the α and β modifications) and $T_{\alpha\beta}$ is the transition temperature. The difference in the molar volumes of the two modifications is usually very small and thus the effect of pressure on the transition point is negligible in a certain range.

1.2. Multicomponent systems

The properties of multicomponent systems are not generally given by the sum of the properties of individual components; the actual contribution of one mole of a component in the mixture to an extensive property of the system is called the partial molar property. If an arbitrary extensive value is denoted by X, then it can be expressed as a function of the temperature, pressure and composition of the system

$$dX = \left(\frac{\partial X}{\partial T}\right)_{P,n_1,n_2,\ldots n_k} dT + \left(\frac{\partial X}{\partial P}\right)_{T,n_1,n_2,\ldots n_k} dP +$$
$$+ \sum_{i=1}^{k} \left(\frac{\partial X}{\partial n_i}\right)_{T,P,n_{j\neq i}} dn_i. \tag{1.15}$$

The partial derivative

$$\bar{X}_i = \left(\frac{\partial X}{\partial n_i}\right)_{T,P,n_{j\neq i}} \tag{1.16}$$

expresses the partial molar value for the ith component in the mixture, so that at constant pressure and temperature eqn. 1.15 can be rewritten in the form

$$dX = \sum_{i=1}^{k} \bar{X}_i\,dn_i, \quad [T,P] \tag{1.17}$$

Integration of eqn. 1.17 for the condition of constant composition leads to

$$X = \sum_{i=1}^{k} \bar{X}_i n_i \quad [T,P] \tag{1.18}$$

If X is a function of state, then it must have a total differential; by comparing the latter with eqn. 1.17 the important relationship among the partial molar values is obtained:

$$\sum_{i=1}^{k} n_i\,d\bar{X}_i = 0. \quad [T,P] \tag{1.19}$$

This is the Gibbs-Duhem equation, which is sometimes written in the form

$$\sum_{i=1}^{k} n_i \frac{\partial \overline{X}_i}{\partial n_j} = 0 \qquad [T, P] \tag{1.20}$$

or

$$\sum_{i=1}^{k} n_i \frac{\partial \overline{X}_j}{\partial n_i} = 0. \qquad [T, P] \tag{1.21}$$

The above relationships are formally simplified for additive values with ideal solutions to give

$$\overline{X}_i = X_i. \qquad [\text{id}] \tag{1.22}$$

Individual phases in a system with s components can exchange various components and therefore the total differentials for the free energies of the individual phases can be written in the form

$$dG' = \left(\frac{\partial G'}{\partial T}\right)_{P, n'} dT + \left(\frac{\partial G'}{\partial P}\right)_{T, n'} dP + \left(\frac{\partial G'}{\partial n'_1}\right)_{T, P, n'_{i \neq 1}} dn'_1 +$$

$$+ \ldots + \left(\frac{\partial G'}{\partial n'_s}\right)_{T, P, n'_{i \neq s}} dn'_s. \tag{1.23}$$

Analogous relationships are valid for the other phases. Since

$$\left(\frac{\partial G'}{\partial n'_i}\right)_{T, P, n'_{j \neq i}} = \mu'_i \tag{1.24}$$

is the chemical potential of the ith component in the first phase, at constant temperature and pressure, eqns. 1.2 and 1.23 can be combined to give

$$dG = \mu'_1 dn'_1 + \mu'_2 dn'_2 + \ldots + \mu'_s dn'_s + \mu''_1 dn''_1 + \ldots +$$
$$+ \mu''_s dn''_s + \ldots + \mu^{(f)}_1 dn^{(f)}_1 + \mu^{(f)}_2 dn^{(f)}_2 + \ldots + \mu^{(f)}_s dn^{(f)}_s = 0. \tag{1.25}$$

Material balance must be maintained, so that

$$dn_1 = dn'_1 + dn''_1 + \ldots + dn^{(f)}_1 = 0 \tag{1.26}$$

with analogous relationships for the other components. The simultaneous validity of the above equations is possible only when

$$\begin{aligned} \mu'_1 = \mu''_1 = \ldots = \mu^{(f)}_1 = \mu_1 \\ \vdots \\ \mu'_s = \mu''_s = \ldots = \mu^{(f)}_s = \mu_s \end{aligned} \qquad [T,P] \tag{1.27}$$

i.e., the chemical potentials of a given component are equal in all of the phases. According to eqn. 1.24, the chemical potential is equal to the partial molar free energy, $\overline{G}_i$, and hence for two-phase equilibrium we have

$$\overline{G}'_1 = \overline{G}''_1 \tag{1.28}$$

so that

$$d\overline{G}'_1 - d\overline{G}''_1 = 0. \tag{1.29}$$

In a system that contains two components, the partial molar free energy of the first component in the first phase is a function of temperature, pressure and the composition of the phase. The total differential then has the form

$$dG_1' = \left(\frac{\partial G_1'}{\partial T}\right)_{P,n} dT + \left(\frac{\partial G_1'}{\partial P}\right)_{T,n} dP + \left(\frac{\partial G_1'}{\partial x_1'}\right)_{T,P,x_{j\neq 1}'} dx_1' \qquad (1.30)$$

where x_1' is the molar fraction of the first component in the first phase. As

$$\mu_i = \mu_i^0 + RT \ln a_i,$$

then

$$d\overline{G}_1' = -\overline{S}_1' dT + \overline{V}_1' dP + RT \, d\ln a_1'. \qquad (1.31)$$

An analogous relationship is valid for the same component in the other phase. On substituting these equations into the equilibrium condition eqn. 1.29, the basic equation

$$-(\overline{S}_1' - \overline{S}_1'') dT + (\overline{V}_1' - \overline{V}_1'') dP + RT \, d\ln \frac{a_1'}{a_1''} = 0 \quad \text{[equil.]} \qquad (1.32)$$

is obtained. When the equilibrium solid phase consists of crystals of pure substance 1, then

$$a_1'' = x_1'' = 1. \qquad (1.33)$$

The effect of various state variables on the solubility in a two-component system can then be shown.

(a) At constant temperature it follows that

$$(\overline{V}_1' - \overline{V}_1'') dP + RT \, d\ln a_1' = 0 \quad [T] \qquad (1.34)$$

i.e.

$$\left(\frac{\partial \ln a_1'}{\partial P}\right)_T = \frac{\overline{V}_1'' - \overline{V}_1'}{RT}. \qquad (1.35)$$

As the difference between the partial molar volumes in the solution and in the solid phase is very small, the effect of pressure on the solubility is negligible.

(b) At constant temperature and pressure, we obtain

$$RT \, d\ln a_1' = 0 \quad [T,P] \qquad (1.36)$$

$$x_1' = \text{const.} \qquad (1.37)$$

The solubility of the substance in a two-component system is constant under these conditions.

(c) At constant pressure the relationship

$$-(\overline{S}_1' - \overline{S}_1'') dT + RT \, d\ln a_1' = 0 \quad [P] \qquad (1.38)$$

or

$$\left(\frac{\partial \ln a_1'}{\partial T}\right)_P = \frac{\overline{S}_1' - \overline{S}_1''}{RT} = \frac{\Delta H_{\text{sol}}}{RT^2} \qquad (1.39)$$

is obtained. For an ideal system this relationship can also be written in the form

$$\left(\frac{\partial \ln x_1'}{\partial (1/T)}\right)_P = \frac{-\Delta H_{sol}}{R}. \quad [\text{id.}] \quad (1.40)$$

The heat of solution, ΔH_{sol}, can be considered to be constant in a narrow temperature range. A graph of log x_1 against $1/T$ is therefore a straight line with a slope of δ, from which the heat of solution can be calculated:

$$\Delta H_{sol} = -19.6\,\delta = 19.6\,\frac{\log (x_1')_{T_2} - \log (x_1')_{T_1}}{\dfrac{1}{T_1} - \dfrac{1}{T_2}}. \quad (1.41)$$

According to eqn. 1.40 and in agreement with the Le Chatelier principle, the solubility of a substance that liberates heat on dissolution decreases with increasing temperature, while the solubility of a substance that absorbs heat from the surroundings on dissolution increases with increasing temperature (the latter condition applies to most substances). If eqn. 1.40 is integrated, assuming that ΔH_{sol} is independent of temperature, the following relationship is obtained

$$\log \frac{(x_1')_{T_1}}{(x_1')_{T_2}} = \frac{\Delta H_{sol}}{2.303R}\left(\frac{1}{T_1} - \frac{1}{T_2}\right) \quad (1.42)$$

corresponding to eqn. 1.41. In an ideal case, the temperature can be extrapolated to the melting point of the substance, thus permitting the ideal solubility to be calculated:

$$\log (x_1')_{T_1} = \frac{\Delta H_{sl}}{2.303R}\left(\frac{1}{T_1} - \frac{1}{T_{sl}}\right) \quad (1.43)$$

where ΔH_{sl} is the latent molar heat of fusion and T_{sl} is the melting point of the pure solute. The use of this equation is, of course, limited to ideal melts, i.e. solutions of very similar substances (isomers, etc.).

It was assumed during integration of eqn. 1.40 that the heat of solution is independent of the temperature. This assumption is, however, valid only in a narrow temperature range; if log x_1 is plotted against $1/T$ over a wider temperature range, a slightly curved line is usually obtained rather than a straight line. Further, if water is used as the solvent, a change in its structure occurs at temperatures between 32 and 34 °C and the solubility curve is usually slightly broken in this temperature range. These drawbacks can be avoided by linearization of the temperature dependence of the solubility by using Othmer's method. For the temperature dependence of the solubility, eqn. 1.40 holds, so that

$$\left(\frac{\partial \ln x_1'}{\partial T}\right)_P = \frac{\Delta H_{sl}}{RT^2}. \quad (1.40)$$

For the temperature dependence of the vapour pressure of the pure solvent, P^0, the formally analogous Clausius-Clapeyron equation holds and can be derived from eqn. 1.12, assuming ideal behaviour of the vapour phase.

$$\left(\frac{\partial \ln P^0}{\partial T}\right)_P = \frac{\Delta H_{\text{vap}}}{RT^2}. \qquad (1.44)$$

By dividing eqns. 1.40 and 1.44, the relationship

$$\left(\frac{\partial \ln x_1'}{\partial \ln P^\circ}\right)_{T,P} = \frac{\Delta H_{\text{sol}}}{\Delta H_{\text{vap}}} = \text{const.} \qquad (1.45)$$

is obtained. Thus, if the corresponding temperatures are correlated with the logarithm of the solvent vapour pressure and log x_1 is plotted against the Othmer temperature scale obtained, straight lines with slopes equal to the ratio of the molar heat of solution and the solvent heat of vaporization are obtained. If a break appears on the straigt lines, a change in the type of solute is unambiguously indicated, e.g. a change of allotropic modification or, in the number of molecules of water of hydration. As the intercept of the linear sections can usually be determined with sufficient precision, a very precise method for the determination of the points of such changes is thus obtained.

Table 1.1 gives suitable temperature scales, corresponding to the $1/T$ values and the Othmer scales for water as the solvent.

So far, we have assumed that the phases behave ideally. More exact relationships are obtained when activities are considered instead of concentrations:

$$\mu_i = \mu_i^0 + RT \ln a_i = \mu_i^0 + RT \ln \gamma_i x_i \qquad (1.46)$$

where a_i is the activity of the ith component, γ_i is the corresponding activity coefficient and μ_i^0 is its standard chemical potential. The phase equilibrium condition eqn. 1.27, which can be rewritten in the form

$$d\mu_1 = d\mu_1' = d\mu_1'' \qquad (1.47)$$
$$d\mu_2 = d\mu_2' = d\mu_2'' \qquad (1.48)$$

serves as the starting point. The total differentials of the chemical potentials of the two components in both phases can be expressed by relationships similar to those given above, i.e.

$$d\mu_1' = -\bar{S}_1' \, dT + \bar{V}_1' \, dP + \left(\frac{\partial \mu_1'}{\partial x_1'}\right)_{T,P} dx_1'. \qquad (1.49)$$

$$d\mu_1'' = -\bar{S}_1'' \, dT + \bar{V}_1'' \, dP + \left(\frac{\partial \mu_1''}{\partial x_1''}\right)_{T,P} dx_1''. \qquad (1.50)$$

Analogous relationships also hold for the other component. The relationship for the first component follows from eqn. 1.47:

$$(\bar{S}_1' - \bar{S}_1'') \, dT - (\bar{V}_1' - \bar{V}_1'') \, dP = \left(\frac{\partial \mu_1'}{\partial x_1'}\right)_{T,P} dx_1' - \left(\frac{\partial \mu_1''}{\partial x_1''}\right)_{T,P} dx_1'' \qquad (1.51)$$

Table 1.1. Coordinates of the scales (in cm) for linearization of the dependence of the solubility on temperature. Scales T−1 and T−2 are calculated using the $1/T$ values, scales O−1 and O−2 are the Othmer temperature scales for water as solvent

t °C	$10^7/T$	T−1	T−2	O−1	O−2
−25	40297	−0.297			
−20	39501	+0.499			
−15	38736	1.264		1.550	
−10	38000	2.000		3.310	
−5	37291	2.709		4.994	
0	36609	3.391		6.608	
5	35950	4.050		8.158	
10	35316	4.684		9.642	
15	34703	5.297	1.485	11.068	0.204
20	34111	5.889	4.445	12.439	4.317
25	33539	6.461	7.305	13.757	8.271
30	32986	7.014	10.070	15.027	12.081
35	32451	7.549	12.745	16.251	15.753
40	31933	8.067	15.335	17.433	19.299
45	31431	8.569	17.845	18.566	22.698
50	30944	9.056	20.280	19.662	25.986
55	30473	9.527	22.635	20.721	29.163
60	30016	9.984	24.920	21.743	32.229
70	29141	10.859	29.295	23.687	
80	28316	11.684		25.503	
90	27536	12.464		27.209	
100	26798	13.202		28.808	
150	23632	16.368		35.528	
200	21134	18.866		40.671	
250	19115	20.885		44.745	
300	17447	22.553		48.092	
350	16047	23.953		50.936	
scale module		1×10^4	5×10^4	10	30

and an analogous relationship is obtained for the other component from eqn. 1.48. These relationships are then multiplied by x_1'' or x_2'' and added. From the Gibbs-Duhem equation, the condition

$$x_1'' \left(\frac{\partial \mu_1''}{\partial x_1''}\right) dx_1'' + x_2'' \left(\frac{\partial \mu_2''}{\partial x_2''}\right) dx_2'' = 0 \qquad [T, P] \qquad (1.52)$$

follows and an analogous relationship for the first phase. By combining eqns. 1.51 and 1.52 the following basic form is obtained:

$$-[x_1'(\bar{S}_1' - \bar{S}_1'') + x_2'(\bar{S}_2' - \bar{S}_2'')] \, dT + [x_1'(\bar{V}_1' - \bar{V}_1'') + x_2'(\bar{V}_2' - \bar{V}_2'')] \, dP +$$
$$+ \frac{x_1'' - x_1'}{x_2''} \left(\frac{\partial \mu_1''}{\partial x_1''}\right)_{T, P} dx_1'' = 0 \qquad (1.53)$$

the relationship for the other phase being analogous. At constant pressure and when $\Delta \bar{H}_i = \bar{H}'_i - \bar{H}''_i = T\bar{S}'_i - T\bar{S}''_i$, this equation can be rearranged to give

$$\left(\frac{\partial x''_1}{\partial T}\right)_P = \frac{x''_2(x'_1 \Delta \bar{H}_1 + x'_2 \Delta \bar{H}_2)}{RT^2(x''_1 - x'_1)\left[\frac{1}{x''_1} + \left(\frac{\partial \ln \gamma''_1}{\partial x''_1}\right)\right]}. \tag{1.54}$$

The equation for the other phase is analogous. After further modification, the following relationship is obtained:

$$R\left(\frac{\partial \ln x''_1}{\partial (1/T)}\right)_P = \frac{x''_2(x'_1 \Delta \bar{H}_1 + x'_2 \Delta \bar{H}_2)}{(x''_2 - x'_2)(1 - \varkappa'')} \tag{1.55}$$

togehter with an analogous equation for the other phase, where

$$\varkappa = \left(\frac{\partial \ln \gamma_1}{\partial \ln x_1}\right)_P = \left(\frac{\partial \ln \gamma_2}{\partial \ln x_2}\right)_P. \tag{1.56}$$

Assuming that only a pure component crystallizes, the above equations can be simplified to give

$$\left(\frac{\partial \ln x_1}{\partial (1/T)}\right)_P = -\frac{\Delta H_{\text{sol}}}{R(1 + \varkappa)} \tag{1.57}$$

differing by the term $(1 + \varkappa)$ from the analogous equation, eqn. 1.40, which holds only for ideal behaviour.

When mixed crystals are formed in the system, $x'' \neq 1$, and eqn. 1.55 must be solved for both phases. The solution is considerably simplified if $\varkappa' \neq \varkappa'' = \varkappa$ because then

$$\begin{aligned}\frac{\partial \ln(x'_1/x''_1)}{\partial (1/T)} &= -\frac{\Delta \bar{H}_1}{R(1 + \varkappa)}. \\ \frac{\partial \ln(x'_2/x''_2)}{\partial (1/T)} &= \frac{\Delta \bar{H}_2}{R(1 + \varkappa)}.\end{aligned} \tag{1.58}$$

For the application of eqn. 1.57 ot 1.58, the concentration dependence of the activity coefficients of the components of the system must be known. These data are usually difficult to obtain. A means of avoiding this difficulty is to carry out calculations by using the simple eqn. 1.40, bearing in mind that ΔH_{sol} in the numerator of the fraction on the right-hand side of the equation does not represent the real value of the heat of solution. The real heat of solution and the temperature dependence of the solubility can be calculated using the nomogram given in Fig. 1.1, in which the appropriate empirical correction is given. A line connecting value 1 on the left-hand scale with temperature T_1 on the right-hand scale intersects a line connecting the c_{T_2}/c_{T_1} value on the left-hand scale with temperature T_2 on the right-hand scale at a point whose position then determines the ΔH_{sol} value on the bottom scale. Another means of avoiding the lack of activity values will be dealt with in the section on correlation of solubility data (p. 146).

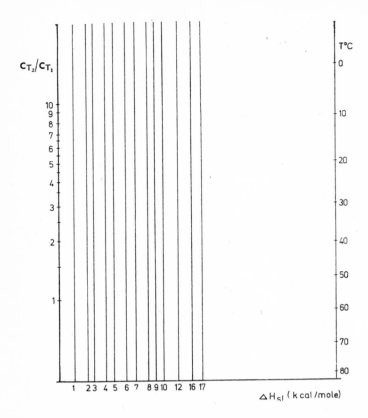

1.1. Nomogram for the calculation of the heat of solution from the temperature dependence of the solubility. (The concentration is expressed in grams per 1000 g of water, temperature in °C and heat in kcal/mole.

1.3. The Gibbs phase rule

Let us consider a closed system with s components that do not react together and that form f phases in equilibrium. For a single phase, $(s-1)$ concentration values would suffice to describe the composition of the system, as the last value can be calculated. Therefore the composition of f phases can be described by a total $f(s-1)$ concentration values. For a complete description of the state of the system, these data must also be further supplemented with the values of the other state variables. For example, if the state of a system is characterized by the temperature and pressure (which are identical in all phases in equilibrium), the number of necessary data must be increased by two values; if pressure exerts a negligible influence, for example, in condensed systems, the data are increased by only one value (the temperature), as the vapour phase is neglected. Hence the total number of data required for the characterization of the composition and state of phases in condensed systems is given by

$$f(s-1)+1.$$

The composition of phases in equilibrium is, however, not independent and is bound by condition 1.27; according to these equations, for the characterization of the chemical potential of an individual component in f phases a single value suffices and is identical with the remaining $(f-1)$ values. For all s components we therefore save a total of $s(f-1)$ values so that the overall number of values necessary for characterization of a condensed system in equilibrium is

$$v = f(s-1) + 1 - s(f-1) = s - f + 1 \tag{1.59}$$

$$v + f = s + 1 \tag{1.60}$$

which is the Gibbs phase law for condensed systems.

List of symbols:

v — the number of degrees of freedom, i.e. the number of independently variable intensive properties that are required for the unambiguous determination of the state of a system at equilibrium;

f — the number of phases, i.e., the number of homogeneous, mutually different parts of the system, physically separated by a phase boundary from the other parts;

s — the number of components, i.e. the smallest number of stoichiometrically independent chemical species by means of which the composition of the system can be characterized. For electrolytes, all actual combinations of ions under the given conditions must be considered as chemical species and it is therefore preferable to determine the number of components as the smallest number of mutually independent ions.

Several examples are given below in order to clarify the type of information that can be obtained from the Gibbs phase rule.

(a) Crystals of calcium nitrate tetrahydrate are in equilibrium with a melt containing the same stoichiometric ratio of calcium nitrate and water as that in the crystals:

$s = 1$ (either the tetrahydrate is assumed to be a compound in both the solid and liquid phases, or two compounds, $(CaNO_3)_2$ and H_2O are considered, which, however, are in identical stoichiometric ratios in both phases and therefore the number of stoichiometric equations, 1, is subtracted from the number of compounds, 2);

$f = 2$ (solid and liquid phases)

$v = 0$.

Hence the state of the system is unambiguously determined by the given conditions (the temperature corresponding to the melting point of calcium nitrate tetrahydrate).

(b) As in (a), but with part of the water evaporated from the melt:

$s = 2$ (the components in the liquid are no longer in a stoichiometric ratio)
$f = 2$ (solid and liquid)
$v = 1$.

For characterization of the system, one more value must be given, either the equilibrium temperature (by which we define the liquid phase composition) or the liquid phase composition (which determines the equilibrium temperature). One degree of freedom corresponds to the solubility curve in phase diagrams.

(c) A cryogenic mixture consisting of sodium chloride, ice and the equilibrium liquid phase:

$s = 2$ (NaCl, H_2O)
$f = 3$ (solid NaCl, solid ice, liquid solution)
$v = 0$.

The composition and the solution temperature in equilibrium are unambiguously determined by the given conditions (the eutectic point).

(d) A system formed by an aqueous solution of sodium nitrate and sodium nitrite, from which part of the sodium nitrite separates in the form of crystals on cooling:

$s = 3$ ($NaNO_2$, $NaNO_3$, H_2O) (or two independent anions + H_2O)
$f = 2$ (liquid, $NaNO_2$ crystals)
$v = 2$.

For characterization of the system, two more independent values must be supplied, i.e., the temperature and, e.g., the sodium nitrate content in the liquid phase. Two degrees of freedom are represented by a plane in the phase diagram or by the solubility curve in the isothermal diagram (T is constant, thus $v = 1$).

(e) The system in example (d), cooled until sodium nitrate crystals also start to separate.

$s = 3$ ($NaNO_2$, $NaNO_3$, H_2O)
$f = 3$ (liquid, solid $NaNO_3$, solid $NaNO_2$)
$v = 1$.

The state of the system is fully characterized by a single value (e.g. temperature). In the isothermal diagram (T is determined so that $v = 0$), the system will be represented by the eutonic point.

(f) The determination of the number of components in a reciprocal salt pair system:

$$KCl + NaNO_3 \rightleftarrows KNO_3 + NaCl \quad [+ H_2O]$$

water + 4 salts − 1 stoichiometric equation
$s = 4$.

2. METHODS OF EXPRESSING THE COMPOSITION OF A SYSTEM

2.1. Solutions and melts

A solution is a homogeneous mixture of two or more components; solutions are either liquid or solid and one component is termed the solute and the other the solvent. The substance that is present in excess and that differs substantially from the solute, for example, in having a considerably lower melting point, is usually denoted as the solvent.

Liquid phases formed by heating a substance or a mixture of substances above their melting points in the absence of a component that might be considered as a solvent are usually called melts.

However, the above definitions are not sufficiently unambiguous to enable a sharp differentiation to be made between solutions and melts. For example, when potassium nitrate is heated to a temperature of 350 °C, i.e., above its melting point, the liquid formed can unambiguously be called a melt. If, however, a small amount of water is added to the melt, the melting point decreases considerably; the system can now be described as a potassium nitrate + water melt or as a solution of water in potassium nitrate (a large excess of nitrate), or as a highly concentrated solution of potassium nitrate in water (water is a common solvent). It is obvious that the first or the last description will most probably be used, as it is unusual to speak about a solution of water in a salt. However, the situation is much less clear when, for example, a phenol—water system is considered, in which two conjugated solutions that differ only in their composition are formed in a certain region; in this instance it is common to speak about a solution of water in phenol or of phenol in water, depending on the predominating component. This ambiguity is also encountered during melting of crystallohydrates, where we can speak about either the melting point (hence melt formation) or dissolution of the salt in its own water of crystallisation (solution formation). We have purposely avoided solid solutions in our discussion; these are actually solidified melts and usually no component in them can be called a solvent (such as a solution of carbon in iron).

Fortunately, this problem concerns only the formal differentiation between two concepts of a liquid phase and it is unimportant from a practical standpoint. Loose definitions of these concepts are sometimes even utilized in order to simplify the task in hand. For example, when the ratio of the concentrations of two or more components is constant in the liquid phase and the content of a single or several other components changes, then the whole system of the former components can be considered to be the solvent (especially when a graphical treatment is employed). In this way it is possible, for example, to study the solubility of ammonium sulphate in a system that contains water, ammonium sulphite and ammonium hydrosulphite at a constant ratio as a simple two-component system of ammonium sulphate and a solvent.

2.2. Concentration units

Different concentration units can be employed in order to describe the composition of phases (especially the liquid phase) in multicomponent systems:

(a) weight per cent: the number of grams of a component in 100 g of solution

$$p_i = 100 g_i / \sum_i g_i$$

(b) weight fraction: the number of grams of a component in 1 g of solution:

$$w_i = p_i/100. \tag{2.2}$$

This is numerically identical with the unit kg/kg.

(c) volume per cent: the number of milliteres of a component in 100 ml of solution:

$$\text{vol--\%} = 100 V_i / V \tag{2.3}$$

(d) volume fraction: the number of millitres of a component in 1 ml of solution:

$$\varphi_i = V_i/V = \frac{g_i/\varrho_i}{\sum g_i/\varrho} \tag{2.4}$$

where ϱ_i and ϱ are the densities of the component and of the solution, respectively.

(e) concentration expressed in terms of grams of a component in 100 g of solvent:

$$a_i = 100 g_i / g_0 \tag{2,5}$$

(f) concentration expressed in terms of grams of a component in 1000 ml of solution:

$$c_i = 1000 g_i / V. \tag{2.6}$$

This is numerically identical with the unit kg/m³.

(g) molar concentration: the number of moles of a component in 1000 ml of solution:

$$C_i = 1000 g_i / (M_i V) = c_i / M_i \tag{2.7}$$

(h) normality: the number of gram equivalents of a component in 1000 ml of solution,

$$N_i = 1000 g_i z_i / (M_i V) = C_i z_i \tag{2.8}$$

where z_i denotes the valence of the ion considered.

(i) molality: the number of moles of a component in 1000 g of solvent:

$$m_i = 1000 g_i / (M_i g_0) \tag{2.9}$$

(j) mole fraction: the number of moles of a component divided by the sum of the number of moles of all of the components in the solution:

$$x_i = n_i / \sum_i n_i \tag{2.10}$$

(k) mole per cent: the mole fraction multiplied by 100:

$$\text{mole} - \% = 100 x_i \tag{2.11}$$

The relationships for the interconversion of the most important units are given in Table 2.1.

Table 2.1. Relationships for conversions among concentration units (ϱ... the solution density, kg/m³; subscripts 1, ..., i, ... s denote solutes, subscript o denotes the solvent)

→	p_i	a_i	c_i	m_i	x_i
p_i	1	$100 p_i / p_0$	$0.01 p_i \varrho$	$\dfrac{1000 p_i}{M_i p_0}$	$\dfrac{p_i / M_i}{\sum_{j=0}^{s} p_j / M_j}$
a_i	$\dfrac{100 a_i}{100 + \sum_{j=1}^{s} a_j}$	1	$\dfrac{a_i \varrho}{100 + \sum_{j=1}^{s} a_j}$	$\dfrac{10 a_i}{M_i}$	$\dfrac{a_i / M_i}{\dfrac{100}{M_0} + \sum_{j=1}^{s} a_j M_j}$
c_i	$\dfrac{10 c_i}{\varrho}$	$\dfrac{100 c_i}{\varrho - \sum_{j=1}^{s} c_j}$	1	$\dfrac{1000 c_i}{M_i (\varrho - \sum_{j=1}^{s} c_j)}$	$\dfrac{c_i / M_i}{\sum_{j=0}^{s} c_j / M_j}$
m_i	$\dfrac{100 m_i M_i}{1000 + \sum_{j=1}^{s} m_j M_j}$	$\dfrac{m_i M_i}{10}$	$\dfrac{m_i M_i \varrho}{1000 + \sum_{j=1}^{s} m_j M_j}$	1	$\dfrac{m_i}{\dfrac{1000}{M_0} + \sum_{j=1}^{s} m_j}$
x_i	$\dfrac{100 x_i M_i}{\sum_{j=0}^{s} x_j M_j}$	$\dfrac{100 x_i M_i}{x_0 M_0}$	$\dfrac{\varrho x_i M_i}{\sum_{j=0}^{s} x_j M_j}$	$\dfrac{1000 x_i}{(\sum_{j=0}^{s} x_j M_j) - x_i M_i}$	1

2.3. Methods of graphical representation

Phase equilibria are often represented graphically, which offers the advantages of first, lucidity, so that the behaviour of the system being studied can be assessed at a glance from the phase diagram, and second, graphical representations can be used directly for calculations of the material balance. The precision of graphical representation depends on the scale employed; even common sizes of graphs permit a precision to be obtained that is sufficient for technical purposes. When phase diagrams are constructed in order to be able to read discrete values, a scale should be selected so that the precision of the reading is comparable to that of the experimental data.

In the following paragraphs, the principal methods of representing various systems are described.

2.3.1. Single-component systems

If the Gibbs phase rule eqn. 1.60 is rewritten for a single-component system, considering the effect of pressure,

$$v + f = 3 \tag{2.12}$$

it is evident that two coordinates, i.e. planar representation, will suffice to describe the behaviour of the system (including the effect of pressure). Temperature is usually plotted on the abscise, pressure on the ordinate. If the system contains a single phase, its geometrical locus includes the whole plane and the system is characterized by two coordinates, the pressure and the temperature. If the pressure and the temperature are varied so that the limits of existence of the phase are not exceeded, no phase change will take place in the system. If the system consists of two phases in equilibrium, there is only one degree of freedom and the locus for such systems is a curve. If the value of one variable, e.g. the temperature, is changed, then the pressure in the system must change so that the point representing the system remains on the phase equilibrium curve; otherwise one of the equilibrium phases disappears. Depending on the type of equilibrium phase, there are lines that correspond to the equilibrium of two solid phases (modifications), of solid and gaseous phases (i.e., lines representing the vapour pressure above the solid phase), of solid and liquid phases (lines representing the effect of pressure on the melting point of the solid) and of liquid and gaseous phases (lines representing the dependence of the vapour pressure on temperature).

If three phases are in equilibrium in a single-component system, there is no remaining degree of freedom,

$$v = 0 \tag{2.13}$$

and the system is represented by the triple point in the phase diagram. This point is characteristic of the given substance and the equilibrium condition determines unambiguously the temperature and pressure corresponding to it. If any of the variables is changed minutely, at least one of the equilibrium phases disappears.

2.3.2. Two-component systems

For two-component systems the Gibbs phase rule assumes the general form

$$v + f = 4. \tag{2.14}$$

Therefore, a three-dimensional diagram is necessary in order to represent such a system when the effect of pressure is taken into account. If, for obvious reasons, we wish to retain a two-dimensional representation, the discussion must be limited to condensed systems thus neglecting the effect of pressure. Then eqn. 1.60 assumes the form

$$v + f = 3. \tag{2.15}$$

The composition and temperature of the system are selected as variables. If the whole phase diagram of a two-component system is plotted, then composition is plotted on the abscissa and temperature on the ordinate. The composition is commonly expressed in weight per cent (or weight fraction) or in mole per cent; the former is encountered more frequently in technical diagrams and the latter in physico-chemical work. Sometimes, only part of the phase diagram is represented; then the independent variable, temperature, is plotted on the abscissa and the dependent variable, concentration, on the ordinate.

If the system contains a single phase, then $v = 2$ and the points representing the system lie on a plane; provided that the limits of the plane are not exceeded, the temperature and the composition of the system can be varied arbitrarily without changing the qualitative characteristics of the system (i.e. without a phase change).

If the system consists of two equilibrium phases (two solids, two liquids or a liquid and a solid), the locus of the representative points is a curve corresponding to one degree of freedom. According to the type of equilibrium, this is a solubility curve (liquid + solid), a limiting miscibility curve for two phases of the same type (liquids or solids) or a curve separating the regions that correspond to the existence of two solids.

If three phases are in equilibrium in a given two-component system, then $v = 0$ and the state of the system is represented by a point. Any change in the state of the system is accompanied by a change in the number of equilibrium phases.

2.3.3. *Three-component systems*

The Gibbs phase rule eqn. 1.60 for a three-component system assumes the form

$$v + f = 4 \tag{2.16}$$

for condensed systems. It is obvious that three-component systems cannot be represented by planar diagrams unless they are simplified. In order to represent phase equilibria in a three-component system in two dimensions, one parameter must be maintained constant (temperature), several different projections must be used simultaneously, or one of the components must be omitted in the representation (e.g. the solvent). These possibilities are discussed below.

A space diagram for a three-component system is depicted schematically in Fig. 2.1. The base of the prism is an equilateral triangle, the apexes of which correspond to the individual components of the system, A, B and C. Temperature is plotted on the ordinate. Points A_0, B_0 and C_0 are the melting points of pure components A, B and C, respectively. The faces of the prism represent the binary systems, A + B, A + C and B + C. Mixtures of substances A and B with compo-

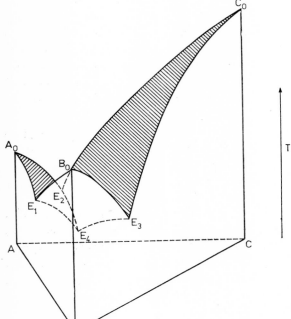

2.1. Scheme of a space diagram of a three-component system.

sition corresponding to point E_1 have the minimum melting point; the other two systems behave analogously. Point E_4 represents the three-component mixture with the lowest possible melting point. A certain temperature, T, is chosen and a horizontal cross-section of the prism is constructed at this temperature. This horizontal plane intersects the planes of the phase diagram in the corresponding isotherms, so that an isothermal phase diagram of a three-component system, represented in a plane and called a triangular diagram, is obtained.

The points representing systems can be plotted on a triangular diagram. An equilateral triangle, ABC (Fig. 2.2), is selected as the coordinate system. The basic equation

$$[A] + [B] + [C] = s \qquad (2.17)$$

holds for the composition of the system, s being one side of the triangle and the brackets denoting concentrations. Side s thus represents the overall content of the components in the system; if the concentrations are expressed in weight per cent, then $s = 100\%$. The composition of the system is then represented by the appropriate points in the triangular diagram.

(a) If the system is represented by an apex of the triangle, e.g., A, then $[B] = 0$, $[C] = 0$ and $[A] = s$ (eqn. 2.17); therefore apex A represents pure substance A and the other two apexes correspond to the other two pure substances.

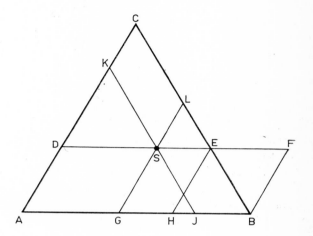

2.2. Representation of a system in a triangular diagram.

(b) If one side of the triangle, e.g. $\overline{AB}$, is the locus of the points representing the system, $[C] = 0$ and hence $[A] + [B] = s$ (eqn. 2.17). A side of the triangle thus represents a two-component system (here A + B) and the composition of such a system is determined by the position of the point, as with binary systems; hence point A corresponds to 100 % of substance A and 0 % of substance B, e.g. 1/4 of the distance from point A to point B lies the point that corresponds to a system containing 75 % of substance A and 25 % of substance B, and point B corresponds to 0 % of substance A and 100 % of substance B. The same holds for the other two binary mixtures, A + C and B + C.

(c) A point inside the triangle (e.g., point S in Fig. 2.2) represents a three-component system. Four geometrical procedures can be used for the determination of the composition of this system:

I. Basic eqn. 2.17 can be rewritten to give

$$\frac{[A]}{s} + \frac{[B]}{s} + \frac{[C]}{s} = 1. \tag{2.18}$$

In Fig. 2.2, straight lines parallel with side $\overline{AB}$ and with the other two sides are drawn through point S. The following relationships then hold:

$$\frac{\overline{DS}}{\overline{DF}} = \frac{\overline{AG}}{\overline{AB}} = \frac{[B]}{s} \tag{2.19}$$

$$\frac{\overline{AD}}{\overline{AC}} = \frac{\overline{GS}}{\overline{AC}} = \frac{\overline{EF}}{\overline{DF}} = \frac{[C]}{s}. \tag{2.20}$$

From the equation

$$\overline{DS} + \overline{SE} + \overline{EF} = \overline{DF} = s \tag{2.21}$$

and from the above relationships it follows that

$$\frac{\overline{SE}}{\overline{DF}} = \frac{[A]}{s}. \tag{2.22}$$

The derivation indicates that, when $s = 100$,

$$[A] = \overline{SE} \, (= \overline{SL})$$
$$[B] = \overline{SD} \, (= \overline{SK})$$
$$[C] = \overline{SG} \, (= \overline{SJ}).$$

Therefore, the content of a component in the system is determined by the length of the section on a straight line parallel with a side of the triangle, drawn from point S away from the apex of the particular component.

II. From the above derivation there follows another method of determining the composition corresponding to point S. If s is again equal to 100 %, then

$$[A] = \overline{SE}$$
$$[B] = \overline{SD}$$
$$[C] = \overline{EF}.$$

Consequently, the contents of the components can be found directly from the sections on lines parallel with the sides of the triangle, drawn through the point representing the system S.

III. The equations

$$\overline{SD} = \overline{AG} \tag{2.23}$$
$$\overline{SE} = \overline{BJ} \tag{2.24}$$
$$\overline{SJ} = \overline{GJ} \tag{2.25}$$

can be written. When eqns. 2.23—2.25 are combined with eqns. 2.19—2.22, then the equations

$$[A] = \overline{BJ}/s \tag{2.26}$$
$$[B] = \overline{AG}/s \tag{2.27}$$
$$[C] = \overline{GJ}/s \tag{2.28}$$

are obtained. Hence lines parallel with the sides of the triangle and drawn through the point representing the system, S, divide the sides of the triangle into three sections, the centre section giving the content of the component represented by the opposite apex of the triangle and each of the other two giving the content of that component whose representative point is at the opposite end of the line.

IV. Other relationships for determining the composition of a system from the position of the point representing it in the triangular diagram can be derived by using Fig. 2.3. The size of the triangle and of the lines drawn from point S at right

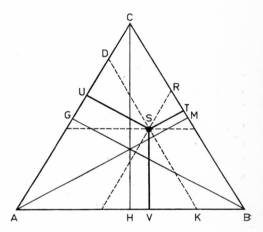

2.3. Alternative method of representing a ternary system.

angles to the sides of the triangle are shown as solid lines, while the dashed lines represent lines parallel with the sides of the triangle. Triangles TSR and CAM are homothetic; hence the equation

$$\frac{\overline{ST}}{\overline{SR}} = \frac{\overline{AM}}{\overline{AC}} \qquad (2.29)$$

is valid and, as

$$\overline{SR} = \overline{CD} \qquad (2.30)$$

then

$$\frac{\overline{ST}}{\overline{AM}} = \frac{\overline{CD}}{\overline{AC}} = \frac{[A]}{s}. \qquad (2.31)$$

From the homothety of triangles SUD and BGC and of SVK and CHB, it can be derived in an analogous manner that

$$\frac{\overline{SU}}{\overline{BG}} = \frac{\overline{RC}}{\overline{BC}} = \frac{[B]}{s} \qquad (2.32)$$

and

$$\frac{\overline{SV}}{\overline{HC}} = \frac{\overline{MB}}{\overline{BC}} = \frac{[C]}{s}. \qquad (2.33)$$

Therefore the lengths of lines drawn from the point representing the system at right-angles to the sides of the triangle are proportional to the concentrations of the components corresponding to the opposite apexes.

Two basic rules follow from the given properties of the triangle:

(a) a line parallel to a side of the triangle is the locus of points representing systems with identical contents of the component corresponding to the opposite apex;

(b) a straight line drawn through an apex of the triangle is the locus of points representing systems with a constant concentration ratio of the two components represented by the remaining two apexes.

There is a habit in technical practice, especially when the solubility of two components in a solvent is described, to use right-angled isosceles triangles for representation of a system of three components instead of equilateral triangles. The main advantage of this method is that special graph paper with triangular coordinate grids is not required and normal graph paper can be used. The plotting of points is also much easier. Fig. 2.4 depicts the representation of systems in rectangular coordinates: system S, consisting of substances A and B and solvent R, contains

$$[A] = \frac{\overline{RT}}{\overline{RA}} \tag{2.34}$$

$$[B] = \frac{\overline{RU}}{\overline{RA}} \tag{2.35}$$

$$[R] = 100 - [A] - [B]. \tag{2.36}$$

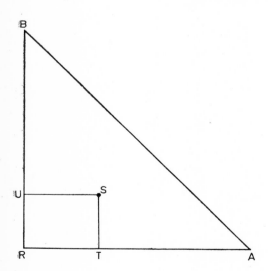

2.4. Right-angled triangular coordinates.

As with two-component systems, the Gibbs phase rule can also be applied to equilibria in three-component systems. If a single phase is present in an isothermal condensed three-component system, it follows from the equation

$$v + f = 3 \tag{2.37}$$

that $v = 2$ and hence a plane is the locus of points representing such systems. If two phases are in equilibrium, for example crystals with a saturated solution, then $v = 1$ and a curve will be the locus of points representing these systems; equilibrium among three phases (e.g., crystals of two components and a saturated solution) is represented by a single point, because $v = 0$. This point is usually called eutonic or invariant and represents a solution saturated with two components.

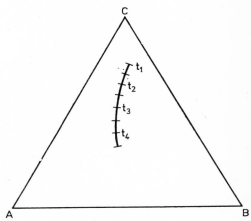

2.5. Temperature dependence of invariant points.

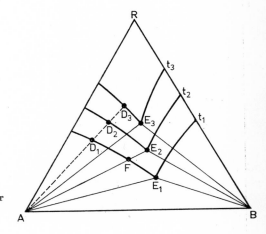

2.6. Various isotherms depicted in triangular diagram.

We have so far discussed only isothermal diagrams of three-component systems, as the representation of temperature as another variable would necessitate a space diagram. This difficulty can be avoided in several ways:

I. Invariant points at various temperatures are plotted on a triangular diagram and are interconnected, so that temperature interpolation can readily be carried out (Fig. 2.5).

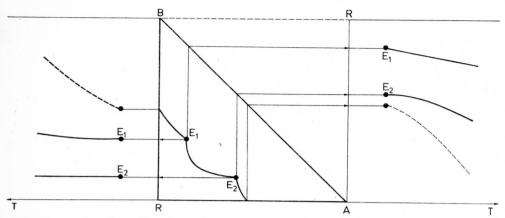

2.7. Temperature dependence in coupled binary diagrams.

II. Individual isothermal diagrams are projected onto a single graph, which then consists of a number of isotherms. This procedure is very advantageous, especially if the graph is to be used for a material balance. An example of such a diagram is given in Fig. 2.6, where the solubilities of substances A and B in solvent R are schematically represented at temperatures t_1, t_2 and t_3, where

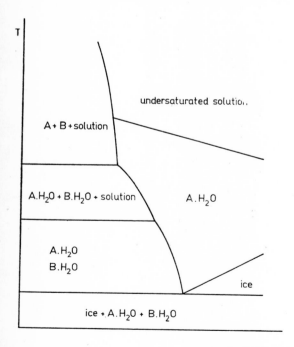

2.8. Aqueous projection of a three-component system.

$t_1 > t_2 > t_3$. For example, when solution D_1 (saturated with substance A at temperature t_1) is cooled down to temperature t_3, crystals of substance A will separate, leaving the mother liquor of composition D_3. Solution F (also saturated with substance A at temperature t_1) can be cooled only to temperature t_2 at which mother liquor E_2 is saturated with both substances (the eutonic point at temperature t_2); on further cooling both substances would separate simultaneously.

III. Isothermal triangular diagrams are supplemented with polythermal binary diagrams, using, for example, the procedure depicted schematically in Fig. 2.7. The position of eutonic points E_1 and E_2 is plotted at various temperatures T using two different projections on the left-hand and right-hand sides of the isothermal traingular diagram. The solubility of the pure components is also sometimes represented (in Fig. 2.7 it is shown as dashed lines).

IV. Three-component systems are sometimes represented on diagrams in order to obtain information on the regions corresponding to the coexistence of various types of phases depending on the temperature. The so-called "aqueous" projection is then employed: temperature is plotted on the ordinate and the composition of the system, expressed in terms of the amount of solvent per 100 units (grams or moles) of the mixture of all solutes (for example, the number of moles of water per 100 moles of all salts), on the abscissa. An example of such a diagram is shown in Fig. 2.8.

V. "Anhydrous" projection is used in a similar manner to "aqucous projection", while differences among the solutes were not considered in the latter projection,

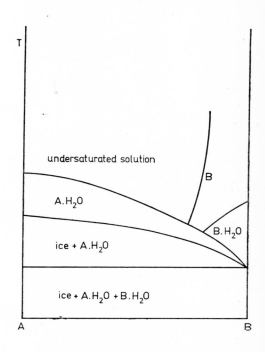

2.9. Anhydrous projection of a three-component system.

"anhydrous" projection supplies these data and therefore both projections are often used for describing systems. A schematic example of "anhydrous" projection is given in Fig. 2.9.

Often, one is not interested in the whole phase diagram; it is then possible to distort the diagram, by a suitable choice of the units plotted on the axes, in order to emphasize a particular concentration range. For example, when the solution region is studied and representation of the system around pure components is not necessary (except for the solvent), the Van't Hoff coordinate system can be used: salt concentrations in grams or moles of salt per 1000 g or 1000 mole of the solvent are plotted on the axes. The points that represent non-aqueous salts are then located at infinity and the regions that correspond to the coexistence of individual phases are limited by parallel lines.

A further method of representing phase equilibria in three-component systems is the Jänecke's projection: the composition of the mixture of solutes in the system (A + B) is plotted on the abscissa, while the water content per 100 mole (or 100 grams) of all of the salts is plotted (analogous to "aqueous" projection) on the ordinate.

All three representation procedures are compared schematically for the NaCl — Na_2SO_4 — H_2O system in Fig. 2.10. The regions in the diagram are denoted as follows: 1, unsaturated solution region; 2, region corresponding to the coexistence of $Na_2SO_4 . 10 H_2O$ and the saturated solution; 3, region corresponding to the coexistence of Na_2SO_4 and the saturated solution; 4, region corresponding to the coexistence of NaCl and the saturated solution; 5, region corresponding to the coexistence of Na_2SO_4, $Na_2SO_4 . 10 H_2O$ and the saturated solution; 6, region corresponding to the coexistence of NaCl, Na_2SO_4 and the saturated solution.

It is evident from Fig. 2.10 that the triangular diagram (centre) provides good overall information on ternary systems, but the regions that correspond to the phase equilibria of the individual salts with the solutions are rather narrow and distortion makes orientation rather difficult. This situation is considerably improved when Van't Hoff projection is employed (bottom); the regions close to the pure anhydrous salts are not shown, but the interesting phase equilibrium region is depicted very clearly. In the Jänecke's projection (top), the point representing water is shifted to infinity and the remaining part of the diagram is shown very clearly, also including part of the region that corresponds to unsaturated solutions. Simultaneously, NaCl is plotted as Na_2Cl_2, in order that the molarities of the common cation be compared on the abscissa.

If certain parts of the diagram are to be emphasized or suppressed, a similar procedure can be employed: an arbitrary multiple of the molecule of a component, whose region is to be emphasized on the diagram is plotted as a component. Fig. 2.11 shows the effect of such coordinate transformation. In the upper part of the figure is depicted a common ternary diagram of system R + A + B with concentrations expressed in mole fractions. In order to construct the diagram at

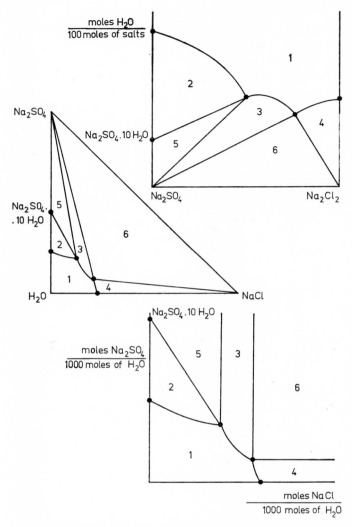

2.10. Comparison of various methods of representing ternary systems.

the bottom of Fig. 2.11, these concentrations were re-calculated for substance A as a trimer (the number of moles of A was divided by three to give the number of moles of A_3). Comparison of the two phase diagrams shows that the region corresponding to the coexistence of A (A_3) with the saturated solutions is increased considerably.

When solid solutions are formed in ternary systems, a procedure similar to that used in representing distillation equilibria is sometimes employed for of the representation data. On a square diagram, one side of which corresponds to

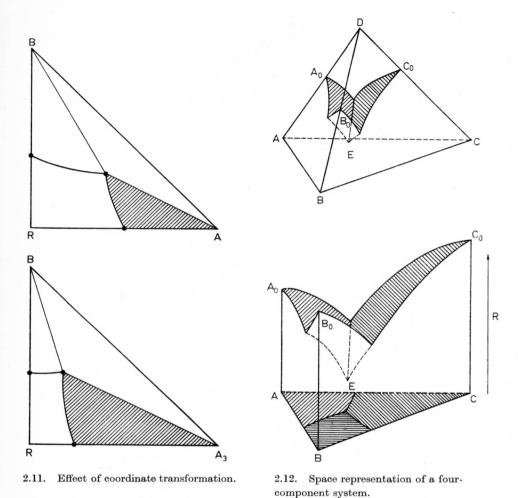

2.11. Effect of coordinate transformation.

2.12. Space representation of a four-component system.

100 wt. —% or 100 mole —%, the content of one of the components in the solid is plotted on one axis and its content in the liquid on the other axis, with respect to the overall solute content.

2.3.4. Four-component systems

The Gibbs phase rule eqn. 1.60 indicates that a four-component condensed system containing a single phase has four degrees of freedom and hence can be represented only in four-dimensional space. When a limitation is set by imposing the condition of constant temperature, a three-dimensional representation becomes possible. An isotherm in a four-component condensed system can be represented by a tetrahedron, if the equation

$$[A] + [B] + [C] + [D] = 100 \% \tag{2.38}$$

is valid, or a trihedral prism with the base formed by the three solutes

$$[A] + [B] + [C] = 100 \%, \tag{2.39}$$

the solvent concentration, [R], being plotted on a suitable scale on the ordinate. Both of these procedures ar depicted schematically in Fig. 2.12. The apexes represent pure components, the edges binary mixtures and the faces ternary mixtures.

A four-component system can be represented two-dimensionally only when two projections are combined, always omitting one of the components; when the solubility of three substances is studied, the Jänecke's "anhydrous" projection, shown schematically at the bottom of Fig. 2.12, often suffices. Then the sides of the triangle represent three-component systems, the apexes two-component systems and the area the four-component system. A typical example of a four-component system is water plus three salts with a common ion. So-called "reciprocal salt pairs" also occur very frequently and are important during conversions of the type

$$AX + BY \rightleftarrows AY + BX \tag{2.40}$$

where A and B represent cations and X and Y anions. Although all four salts can coexist in aqueous solution, the composition of the solution can be specified in terms of the contents of three salts and water, as the fifth concentration value can be calculated from the stoichiometry of eqn. 2.40. The isotherm of such a system can again be represented in space, either as a pyramid with a square base or (more frequently) as a tetrahedral prism (Fig. 2.12). The apexes of the prism represent pure salts AX, AY, BX and BY and the vertical edges represent solutions of these salts in solvent R. The prism faces represent the corresponding three-component systems of two common ion salts plus solvent type. In a similar manner to the example discussed above Jänecke's "anhydrous" projection permits two-dimensional representation of a reciprocal salt pair system. This projection is represented schematically in the upper part of Fig. 2.13 on the base of the tetrahedral prism. The apexes of the square represent solutions of the pure salts, the sides correspond to ternary systems and the area of the square to the four-component system. The area is subdivided into four parts, separated by curves. The areas represent equilibria between the salts and saturated solutions (I, AX; II, AY; III, BY; IV, BX). The curves separating the areas are the loci of solutions saturated with two salts ($\overline{SW}$, AX + AY; $\overline{TW}$, AY + BY; $\overline{WZ}$, AX + BY; $\overline{UZ}$, BY + BX; $\overline{VZ}$, BX + AX; no solution can be saturated with AY and BX simultaneously) and points W and Z represent solutions saturated with three salts simultaneously (AX + AY + BY and AX + BX + BY, respectively).

The bottom part of Fig. 2.13 depicts the plotting and reading of points in the Jänecke's projection. The basis of the system is 100 mole of the salt mixture, the composition of which can be expressed in terms of three salts. For example, the mixture contains:

Salt	Moles of salt	Number of gram-ions			
		A	B	X	Y
AX	20	20	—	20	—
BX	30	—	30	30	—
BY	50	—	50	—	50
Σ	100	20 100	80	50 100	50

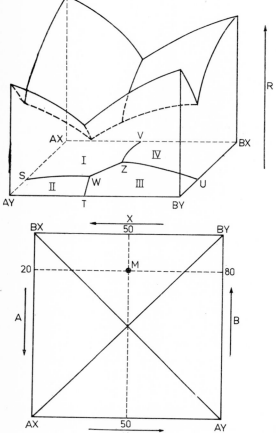

2.13. Representation of a reciprocal salt pair system.

This system is plotted as point M in Fig. 2.13. The diagonals divide the square into two triangles. Point M is located in triangle AX-BX-BY, which corresponds to the specification of the composition of the system. However, it also lies in the triangle BX-BY-AY, so that its composition could also be expressed in terms of the following three salts:

Salt	Number of gram-ions				Moles of salt
	A	B	X	Y	
sum of all salts	20	80	50	50	100
		100		100	
BX	—	50	50	—	50
BY	—	30	—	30	30
AY	20	—	—	20	20

In both typical examples of four-component systems, "anhydrous" projection was necessary in order to be able to represent the system in a plane. For representation of the content of the fourth component (the solvent), the Jänecke's projection must be complemented by another projection that involves this component. Examples of the complete representation of a four-component system are given in Fig. 2.14.

System $A + B + C + D$ is represented in the upper half of Fig. 2.14; the basic representation is the Jänecke's projection from apex D of the tetrahedron to the opposite face, thus obtaining the "anhydrous" projection A-B-C, which is represented in rectangular coordinates (the triangle shown in bold lines). This basic projection is complemented by two "aqueous" projections, one of which suffices to describe the system. In the left-hand part, the projection of the content of the fourth component at individual points in the system is depicted; the straight line AC represents zero content of component D and apex D corresponds to pure component D. To the right of the triangle another means of representing the content of component D is shown, using the above-mentioned coordinates, grams or moles of D per 100 g or 100 mole of the sum of substances $A + B + C$.

The bottom part of Fig. 2.14 shows a complete projection of a reciprocal salt pair system. The basic diagram is again formed by the Jänecke's "anhydrous" projection, which is complemented by an "aqueous" projection in coordinates of moles of water per 100 mole of the sum of all salts; the composition of the system is characterized on the abscissa in terms of the ratio of X and Y anions in the system.

The effect of temperature in four-component systems can be expressed by superimposing two or more isothermal diagrams, as in three-component systems.

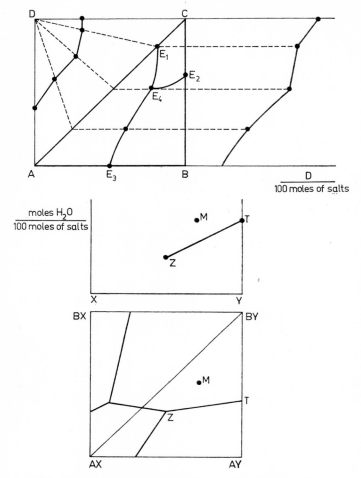

2.14. Complete representation of a four-component system.

2.3.5 Multicomponent systems

The graphical representation of multicomponent systems is usually very complicated. Basically, the representation can be converted into a series of two-dimensional diagrams by representing the multicomponent system as a complex geometrical body in three-dimensional space and by projecting the isotherms that correspond to the phase equilibria on a plane in various ways. The projection procedure shown in the upper left-hand part of Fig. 2.14 for a four-component system is frequently employed.

Another procedure that can also be derived for the projection of a suitably selected space body depends on keeping the amount of $(s-3)$ or $(s-4)$ components of an *s*-component system constant and representing the system as a three- or four-component system, one of the components corresponding to the sum of the components that are kept constant.

With an increasing number of components, the diagrams become very complicated and are not very useful in practice for representing the system.

2.3.6. *The straight line rule and the lever rule*

It has already been mentioned in the previous sections that phase diagrams can be used not only for qualitative but also for quantitative descriptions of systems and that they can form a basis for calculating material balances. In order to be able to utilize this property, the quantitative relationships among the individual phases in equilibrium most, of course, be known. These quantitative relationships are represented graphically by two rules, the straight-line rule and the lever rule.

An *s*-component system with weight g and composition $p_1, p_2, p_3, \ldots, p_s$ can be considered in general terms. This system is divided into two phases, ' and ", the first of which has overall weight g' and composition $p'_1, p'_2, \ldots, p'_s$. The weight and composition of the other phase can then be determined. From the material balance, it follows that

$$g'' = g - g' \tag{2.41}$$

and, for the balance of an arbitrary *i*th component,

$$p''_i g'' = p_i g - p'_i g'. \tag{2.42}$$

The overall weight of the system, g, expressed by eqn. 2.41, is substituted into eqn. 2.42, giving the equation

$$p''_i g'' = p_i g' + p_i g'' - p'_i g'. \tag{2.43}$$

This equation can then be modified to give

$$\frac{g''}{g'} = \frac{p'_i - p_i}{p_i - p''_i}. \tag{2.44}$$

This relationship holds in general for the *i*th component. Therefore, it follows for individual components of the system that

$$\frac{p'_1 - p_1}{p_1 - p''_1} = \frac{p'_2 - p_2}{p_2 - p''_2} = \ldots = \frac{p'_s - p_s}{p_s - p''_s}. \tag{2.45}$$

Eqn. 2.45 expresses the condition that all three points (representing the initial system and the two phases) lie on a single straight line. Therefore, the straight line rule states that the point representing the system and those for the phases that comprise it are located on a single straight line.

Note: The condition that the phases must be in equilibrium was not imposed during the derivation; therefore, the rule is general.

On inspection of the graphical plot of eqn. 2.44 in Fig. 2.15, it is found that the weight ratios of the individual conjugated phases are inversely proportional to the sections on the line that connects the points representing the phases. As this

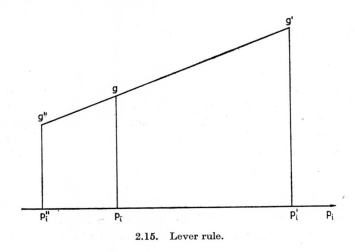

2.15. Lever rule.

formulation is reminiscent of the force ratio for a lever employed in physics, eqn. 2.44 is known as the lever rule. The lever rule is sometimes used in a modified form:

$$\frac{g''}{g} = \frac{p'_i - p_i}{p'_i - p''_i} \, . \tag{2.46}$$

2.3.7. *Enthalpy diagrams*

So far, only those phase diagrams which permitted monitoring of the effect of temperature in addition to the composition of the system have been discussed. Representation of this quantity was very easy and lucid, especially with two-component systems, for which the temperature was plotted against the composition of the system. If another quantity, the enthalpy (heat content), is plotted in an analogous manner against the composition of a two-component system, a diagram is obtained, that permits the thermal balance for the processes that take place in the given system to be described, in addition to the material balance.

A typical enthalpy diagram is shown in Fig. 2.16 for the $Na_2B_4O_7 - H_2O$ system.[1] The bold curve on the left-hand part of the diagram represents saturated

---

[1] S. Scholle: Chem. Průmysl *15*, 530 (1965).

2.16. Enthalpy diagram in the Na₂B₄O₇—H₂O system.

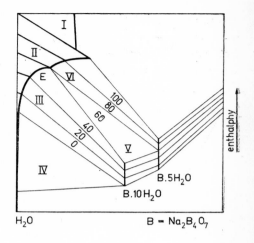

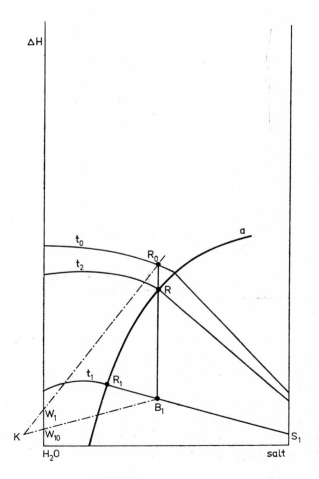

2.17. Enthalpy diagram for the cooling process.

solutions; to the left of it, the isotherms of unsaturated solutions for the temperature range 0—100 °C are located in region *II* and above them, in area *I*, is the heterogeneous liquid—vapour region. To the right of the saturated solution curve lie the heterogeneous region *III*, corresponding to equilibrium between

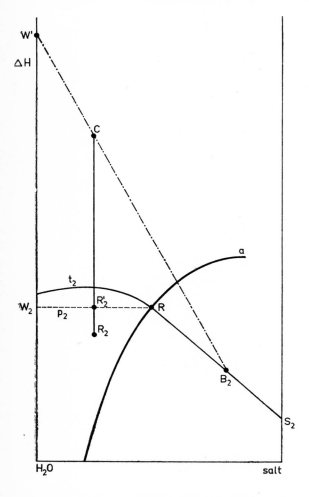

2.18. Enthalpy diagram for isothermal evaporation.

crystals of the decahydrate and the saturated solutions, and region *IV*, corresponding to equilibrium between crystals of the pentahydrate and the saturated solutions. Region *V* represents equilibrium among the crystals of the pentahydrate and the decahydrate and solution E saturated at 60.8 °C. Region *VI* corresponds to coexistence of the decahydrate and ice.

The cooling process in the enthalpy diagram is depicted schematically in Fig. 2.17. Solution R_0, with an initial temperature of t_0, is cooled down to temperature t_1. At temperature t_2 (point R), the crystals of substance S start to separate, crystallisation progresses and, finally, when the system is cooled to point B_1 at temperature t_1, crystals S_1 and mother liquor R_1 are obtained in the weight ratio given by the lever rule, $R_1 : S_1 = \overline{S_1B_1} : \overline{R_1B_1}$. Segment $\overline{R_0B_1}$ represents the amount of heat that must be removed from solution R_0. If cooling water with an

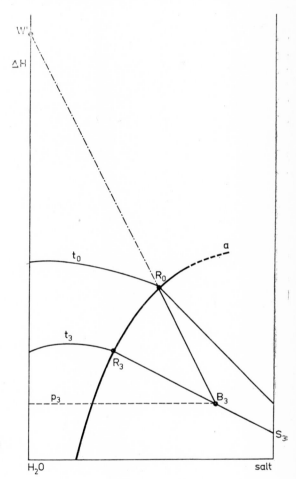

2.19. Enthalpy diagram for adiabatic evaporation.

enthalpy corresponding to W_{10} is used and it is assumed that its enthalpy increases to a value corresponding to point W_1, then the amount of cooling water consumed will be related to the amount of solution according to the ratio of the segments, $\overline{R_0K} : \overline{W_1K}$ or $\overline{B_1K} : \overline{W_{10}K}$.

Fig. 2.18 depicts schematically the process of isothermal evaporation in an enthalpy diagram. Unsaturated solution R_2 is heated to its boiling point at pressure p_2; water W_2 evaporates until the concentration of the saturated solution, R, is obtained. The ratio of the amount of water evaporated to the amount of saturated solution equals the ratio of the segments, $\overline{R_2'R} : \overline{W_2R_2'}$. Further, the evaporation is isothermal at temperature t_2 until the required suspension is obtained, for example, B_2, when the weight ratio of crystals S_2 and mother liquor R equals the ratio of the segments, $\overline{RB_2} : \overline{B_2S_2}$. The overall amount of heat supplied to the solution is $\overline{R_2C}$ and point W' represents the enthalpy of water vapour.

Fig. 2.19 depicts schematically the process of adiabatic evaporation in an enthalpy diagram. Solution R_0, saturated at temperature t_0, is injected into the crystalliser in which a decreased pressure, p_3, is maintained. By evaporation of water (the enthalpy of the water vapour is given by point W'), the solution injected is thickened and cooled down to temperature t_3, corresponding to point B_3. The weight ratio of crystals S_3 and mother liquor R_3 is $\overline{R_3B_3} : \overline{B_3S_3}$ and the ratio of the amount of water evaporated to the suspension is $\overline{B_3R_0} : \overline{R_0W'}$.

It is evident that enthalpy diagrams are very useful for solving many technical problems. However, their construction is very tedious and so far only a few have been published.

3. TYPES OF SYSTEMS AND THEIR PHASE DIAGRAMS

Various types of systems are surveyed in Table 3.1. The classification depends on the number of components, their properties and their miscibilities. The number

Table 3.1. Classification of *s-l* heterogeneous systems

Number of components	Class	Group		Properties
1	I			a single modification
		a		triple point below atmospheric pressure
		b		triple point above atmospheric pressure
	II			several modifications
		a		all modifications stable
		b		one modification unstable
2	I			components immiscible in the solid phase
		a		components do not form a stoichiometric compound
			a_1	components completely miscible in the liquid phase
			a_2	components partially miscible in the liquid phase
			a_3	components immiscible in the liquid phase
		b		components form a stoichiometric compound
			b_1	a compound stable up to its m.p.
			b_2	a compound unstable at its m.p.
	II			components completely miscible in the solid phase
		a		components do not form a solid compound
			a_1	melting point curve without a maximum or minimum
			a_2	melting point curve exhibits a maximum or minimum
		b		components form a solid compound
	III			components partially miscible in the solid phase
		a		close melting points of the components
		b		melting points of the components far apart
3	I			components immiscible in the solid phase
		a		components do not form a stoichiometric compound
		b		components form a stoichiometric compound
			b_1	binary compounds formed
			b_2	ternary compounds formed
	II			components miscible in the solid phase
		a		components completely miscible in the solid phase
		b		components partially miscible in the solid phase
4	I			four different components
	II			reciprocal salt pairs

of possible combinations increases with the number of components; therefore, only the most important types are listed for three- and four-component systems.

In subsequent sections we then consider the phase diagrams for the individual systems. As phase diagrams are most readily understood and their properties are best explained when examples are given, the behaviour of characteristic systems on cooling or solvent evaporation are described for each diagram.

3.1. Single-component systems

Type 1-Ia: a single modification, triple point below atmospheric pressure (Fig. 3.1, top).

The diagram is divided into three regions by three curves, corresponding to the solid (s), liquid (l) and gaseous (g) phases. If the liquid, whose state (temperature T and vapour pressure P) is given by point A, is cooled, its pressure decreases and

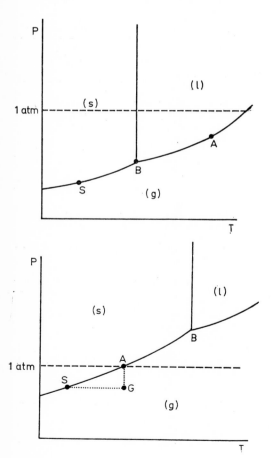

3.1. Phase diagram of 1-Ia, b system type.

the state of the system follows the liquid-vapour equilibrium curve AB to point B. This is the triple point, at which the three phases are in equilibrium. On further removal of heat from the system, crystals will separate from the liquid phase at a constant temperature until the substance solidifies. Then the temperature starts to decrease again and the pressure simultaneously decreases along the sublimation curve, BS, representing the vapour pressure above the solid. The almost vertical curve starting at point B is the fusion curve; the dependence of the melting point on the pressure is almost negligible, as follows from eqn. 1.21. Direct transfer from the gaseous to the solid phase can occur only at temperatures below the triple point and thus at relatively low partial pressures.

Type 1-Ib: a single modification, triple point above atmospheric pressure (Fig. 3.1. bottom)

This type of phase diagram is formally identical with the previous one, differing only in that the triple point is located above atmospheric pressure. Therefore solid-liquid transfer cannot be carried out under atmospheric pressure; the solid evaporates at a temperature corresponding to the vapour pressure equal to atmospheric pressure (point A), the vapours are mixed with inert substances (segment $\overline{AG}$) and, on cooling, the component can again be condensed from the gaseous mixture directly to the solid phase (point S). The whole process is termed sublimation and desublimation.

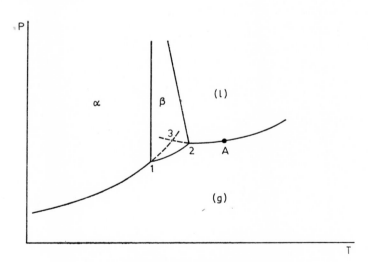

3.2. Phase diagram of 1-IIa system type.

Type 1-IIa: all modifications are stable (Fig. 3.2)

The substance occurs in two modifications, α and β. Thus, α denotes the region corresponding to the existence of stable modification α in the diagram, β is the region corresponding to stable modification β, l is the melt region and g is the region corresponding to the vapour phase. The curve running from the left to point *1* represents phase equilibrium between solid modification α and the gaseous phase, the section of the curve between points *1* and *2* is the sublimation curve of solid modification β and the curve from point *2* to the right corresponds to the vapour pressure over the liquid. Point *1* is the triple point, corresponding to equilibrium among the three phases consisting of modifications α and β and the gaseous phase. From this point runs an almost vertical curve, representing the dependence of the point corresponding to the change of modification α into modification β on the pressure. Similarly, point *2* is the triple point corresponding to equilibrium among the solid modification β, the melt and the gaseous phase. The almost vertical curve starting here represents the effect of pressure on the melting point of modification β. According to eqns. 1.21 and 1.22, the latter two curves are almost vertical but their slopes are slightly different. If these curves intersect (at very high pressures in this diagram), the system will behave as a 1-IIb system at pressures above the intercept and as a 1-IIa system at lower pressures.

If the melt represented by point A is cooled slowly, the point representing the system moves along the pressure curve to the left until stable modification β begins to separate at point *2*. Then the temperature remains constant on further removal of heat, corresponding to point *2*, until the melt solidifies; the heat removed is compensated for by the latent heat of solidification liberated. When all of the solid has separated in the form β, the temperature begins to decrease again on further cooling until the point representing the system moves to point *1*. At this point, solid modification β begins to be converted into modification α, again at a constant temperature. The heat removed is compensated for by the latent heat of change of modification. A further decrease in the temperature takes place only after all of modification β has disappeared.

Heating of solid modification α leads to the reverse process: at point *1* modification α changes into β at a constant temperature and at point *2* modification β melts at a constant temperature. Changes of substances that exist in two stable modifications and for which the change in modification is reversible are termed enantiotropic.

If the melt given by point A is cooled very rapidly, it may happen that solid phase β does not begin to separate at point *2* and the melt is supercooled. If the supercooling progresses to point *3*, modification α can separate directly; it is metastable in this region and changes slowly and spontaneously into modification β, unless temperatures below point *1* are attained rapidly.

Sulphur is a typical example of a system that undergoes an enantiotropic change.

Type 1-IIb: one modification is unstable (Fig. 3.3).

This type of diagram is typical of substances that occur in one stable and one metastable modification. In Fig. 3.3, α is the region corresponding to the existence of stable modification α, the part above the dashed line corresponding to metastable modification β. The melt region is denoted as l and that of the gaseous

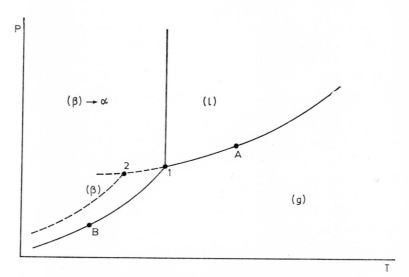

3.3. Phase diagram of 1-IIb system type.

phase as g. The solid line from the left up to point *1* corresponds to equilibrium between stable modification α and the gaseous phase. The dashed curve represents metastable equilibrium between unstable modification β and the gaseous phase. The line from point *1* running upwards represents the effect of pressure on the melting point of modification α. Point *1* is the triple point for equilibrium among modification α, the melt and the gaseous phase. Point *2* corresponds to metastable equilibrium among unstable modification β, the melt and the gaseous phase. The curve from point *1* to the right represents the vapour pressure above the melt.

If the melt corresponding to point A is cooled, then the point representing the system will shift to point *1*. Here stable modification α should begin to separate; this actually happens when the cooling is very slow or when the melt is seeded with a crystal of modification α at the temperature given by point *1*. However, on faster cooling the melt will be supercooled until metastable modification β begins to separate at point *2*; this can then spontaneously recrystallise over a long period of time to give stable modification α. It is interesting that the metastable modification usually separates before the stable modification (the Ostwald rule); This phenomenon is caused by the lower consumption of energy for rearrangement of the particles of the melt to give modification β, compared with modification α. If, on the other hand, the crystals of modification α heated, the state of the system proceeds from, e.g., point B to point 1, where all of the crystals fuse at a constant temperature and only then does the temperature of the melt begin to increase again. Therefore it is impossible to proceed directly from stable modification α to metastable modification β; in order to achieve this conversion, modification α must be fused and the melt supercooled so that metastable modification β can separate. The conversion of the β into the α modification is, of course, spontaneous. Conversions in systems in which a change in modification can proceed in only one direction are called monotropic changes.

A typical example of a monotropic system is phosphorus, with stable red and metastable white modifications.

3.2. Two-component systems

Type 2-Ia$_1$: the components do not form a compound and are completely miscible in the liquid phase (Fig. 3.4).

A phase diagram of two substances R and S (where R can denote a solvent) is depicted in Fig. 3.4. Area I represents the unsaturated solution, area II corres-

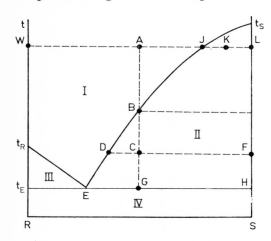

3.4. Phase diagram of 2-Ia$_1$ system type.

ponds to the coexistence of crystals of substance S and the saturated solution, area III represents the coexistence of crystals of substance R and the saturated solution and area IV corresponds to mixtures of crystals of substances R and S. The curve separating regions I and II is the solubility curve for substance S, while that separating regions I and III is the curve for phase equilibrium between solid R and the corresponding liquid. The curves intersect at point E, where solid R, solid S and a solution with composition E, saturated with both substances, are in equilibrium. Points t_R and t_S are the melting points of pure components R and S, respectively.

If an unsaturated solution, represented by point A, is cooled, the composition of the system does not change and the point representing the system therefore moves vertically. This vertical line intersects the solubility curve at point B, lying on the boundary of the region corresponding to the separation of crystals of substance S. On further cooling, crystals of substance S separate, the mother liquor is depleted in this component and hence its composition varies along the solubility curve from right to left. For example, on cooling the system down to a temperature corresponding to point C, crystals of composition F and the mother liquor (melt) with a composition corresponding to point D are in equilibrium in the weight ratio $\overline{CD} : \overline{CF}$. On a further decrease in temperature, the point representing the liquid phase moves along the solubility curve towards point E; finally, at a temperature corresponding to point G, crystals of H are in equilibrium with mother liquor of composition E. Solution E is saturated with both components, so that the crystals of both components will separate from a liquid phase with a constant composition at constant temperature t_E on further removal of heat, until the entire liquid phase disappears. Temperature t_E is thus the lowest temperature at which crystals of a single component can still be obtained from the solution. For initial solution A, the weight ratio of the maximum obtainable amount of crystals of S to mother liquor E is given by the ratio of segments $\overline{EG} : \overline{GH}$. Point E is called the eutectic point, temperature t_E is the eutectic temperature and the mixture of substances with composition corresponding to point E is a eutectic mixture.

Note: The eutectic temperature corresponds to phase equilibrium of two solids and one liquid. If substances R (e.g.) ice) and S (e.g., rock salt) are mixed in a ratio corresponding to the eutectic mixture and the equilibrium liquid phase is formed, the temperature of the system stabilizes at the t_E value until the solid phase disappears. As $t_E < t_R$, this effect is utilized in the preparation of freezing mixtures.

In an analogous manner, on isothermal evaporation of solvent R from solution A, the temperature remains constant and the point representing the system moves to the right. At point J, the solution becomes saturated (the ratio of the amount of water evaporated to the initial weight of solution A equals the ratio of the segments, $\overline{AJ} : \overline{WJ}$) and the crystals of substance S begin to separate. The point representing the system moves further to the right so that the ratio of the weight of crystals L to that of the mother liquor J, for example at point K, is $\overline{KJ} : \overline{KL}$, and the ratio of all of the water evaporated to the initial amount of solution A is $\overline{AK} : \overline{KW}$. Finally, all of solvent R is evaporated from the solution at point L and only solid S remains.

Typical examples of 2-Ia_1 systems are potassium chloride-water, sodium sulphate — sodium chloride, potassium chloride — silver chloride and bismuth-cadmium.

Type 2-Ia$_2$: the components do not form a compound and are partially miscible in the liquid phase (Fig. 3.5).

If a solution whose composition and temperature are characterized by point C is cooled, a new liquid phase with composition given by point F appears on reaching point D. This point represents a solution of substance B in substance A, whereas point F corresponds to a solution of substance A in substance B. The two solutions are termed conjugated; they are characterized by identical activities (or chemical potentials) of the dissolved component. During further cooling, the compositions of the two conjugated phases vary along curves DG and FH.

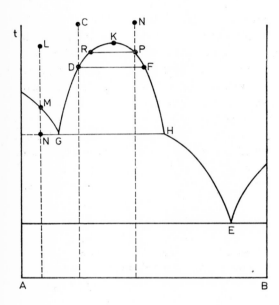

3.5. Phase diagram of 2-Ia$_2$ system type.

This separation into two liquid phases can generally take place only at temperatures below the critical solution temperature, the value of which is given by the coordinate of point K (the critical dissolution point). As soon as the temperature decreases to the value given by points G and H, solid A begins to separate and the temperature remains constant until one of the conjugated liquid phases disappears (G). On further cooling, the temperature again decreases, substance A separates and the composition of the solution varies from point H up to the eutectic point, E.

A similar process takes place when a solution of composition N is cooled. The conjugated liquid phase, R, appears at point P, liquid phase G disappears at the temperature corresponding to points G and H and component A then separates normally.

On cooling solution L, solid phase A begins to separate at point M and the composition of the solution gradually changes up to point G. Then substance A separates at a constant temperature, liquid phase H appears in addition to the disappearing liquid, G, and increases in amount at the expense of phase G. After liquid phase G has disappeared, the temperature again begins to decrease and the composition of the solution varies along curve HE.

Benzoic acid-water and phenol-water systems are typical examples of the 2-Ia$_2$ type of two-component system.

Type 2-Ia$_3$: the components do not form a compound and are immiscible in the liquid phase (Fig. 3.6).

In systems in which the two components are immiscible in both the solid and liquid phases, the components do not affect one another. Therefore, first one compound fuses at its melting point (at a constant temperature) and then the

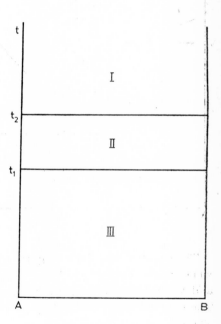

3.6. Phase diagram of 2-Ia$_3$ system type.

other melts at a higher temperature. Area I in the diagram represents the coexistence of two immiscible liquids, area II is the region corresponding to the coexistence of one pure liquid component (melting point t_1) and one solid (melting point t_2) and area III represents the existence of two separate solids.

Examples of this type of system are silver — vanadium and bismuth — chromium.

Type 2-Ib$_1$: the components form a compound stable up to the melting point (Fig. 3.7).

If the components of the system form a compound with a constant melting point, the phase diagram can be divided into two 2-Ia$_1$ type diagrams, corresponding to systems A — A$_x$B$_y$ and A$_x$B$_y$ — B. Point E_1 is the eutectic point of the former system, point E_2 is that of the latter. Point M corresponds to the melting point of binary compound A$_x$B$_y$: the compound melts at a constant temperature and the melt has the same stoichiometric composition; such compounds are said to have congruent melting points.

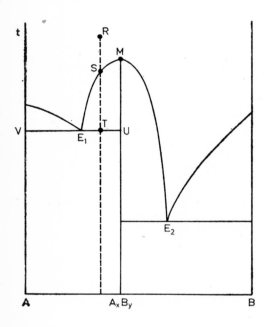

3.7. Phase diagram of 2-Ib$_1$ system type.

The explanation of the diagram resembles that for the 2-Ia$_1$ type. For example, when saturated solution R is cooled, substance A$_x$B$_y$ begins to separate at point S and the process continues down to the eutectic temperature, given by point T. The composition of the solution varies simultaneously from point S to point E_1. The ratio of the weight of the crystals, U, to that of the mother liquor equals the ratio of the segments, $\overline{E_1T} : \overline{TU}$. On further cooling, substances A$_x$B$_y$ (point U) and A (point V) will separate simultaneously until the entire liquid phase E_1 disappears.

Examples of this type of system are calcium chloride-potassium chloride, sodium fluoride — sodium sulphate and water-sulphuric acid. In the last system, a series of compounds are formed, so that the phase diagram exhibits a number of maxima and can be separated into a number of partial systems.

Type 2-Ib$_2$: the components form a compound unstable at the melting point (Fig. 3.8).

In this system, the components form a compound that decomposes below its melting point, which on further heating forms a melt with a composition different from that corresponding to the stoichiometry of the compound. These compounds are said to have incongruent melting points. A typical phase diagram for this type of system is shown in Fig. 3.8. The individual areas represent the following: *I*, the saturated solution; *II*, the coexistence of solid B and saturated solutions with compositions given by the curve *J-C-P*; *III*, the coexistence of solid A$_x$B$_y$

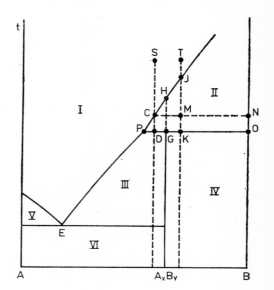

3.8. Phase diagram of 2-Ib$_2$ system type.

and saturated solutions with compositions given by the curve *P-E*; IV, the existence of heterogeneous mixture of solids A$_x$B$_y$ and B; *V*, the coexistence of solid A and the corresponding saturated solutions; and *VI*, the existence of heterogeneous mixtures of solids A and A$_x$B$_y$. Point *E* is the eutectic point of the A — A$_x$B$_y$ system, while point *P* is the peritectic point; at both the eutectic a peritectic temperatures, the phase change proceeds at a constant temperature.

The starting point is *S*, which represents the composition of a system that is richer in component A than corresponds to the stoichiometry of the compound. On cooling, crystals of substance B begin to separate at point *C*; this process progresses until the temperature decreases to the peritectic value at point *D*. Meanwhile, the equilibrium composition of the solution changes from point *C* to point *P*. Therefore, at point *D*, a mixture of crystals *O* and solution *P* is obtained, in a weight ratio equal to the ratio of the segments, $\overline{PD} : \overline{DC}$. On removal of further heat, the temperature of the system does not change, the amount of crystals of B will decrease and crystals of substance A_xB_y will begin to be formed. The point representing the solid phase will shift from O to G. At the end of this peritectic recrystallisation a mixture of crystals A_xB_y and of saturated solution P is obtained, in the weight ratio $\overline{PD} : \overline{DG}$; the increase in the mass of the crystals corresponds to the stoichiometric amount of component A and part of component B and, simultaneously, the amount of solution with composition P decreases. At the instant when solid B disappears, the temperature begins to decrease again and substance A_xB_y separates on further cooling, the composition of the solution varying along curve PE.

Heating of substance A_xB_y proceeds in an analogous manner in the opposite direction. The crystals can be heated to the incongruent melting point, G, at which they decompose to give crystals of substance B (point O) and peritectic solution P in the weight ratio, $\overline{GP} : \overline{GO}$; on further heating, crystals of B dissolve and disappear completely at the temperature corresponding to point H.

The behaviour of system T, which has a higher content of B than is indicated by the stoichiometry of A_xB_y, is different. On cooling, crystals of substance B again start to separate at point J, so that the ratio of the weight of crystals N to that of the mother liquor C is $\overline{CM} : \overline{MN}$ at point M. Crystals of substance B separate down to the peritectic temperature at point K, at which temperature a mixture of crystals of substance B (point O) and solution P is obtained. On removal of further heat, crystals of substance B separate at a constant temperature, and simultaneously the amount of solution P decreases. The liquid mixture, P, disappears completely at the end of the peritectic recrystallization and a heterogeneous mixture of crystals of substances B and A_xB_y is obtained in the weight ratio, $\overline{GK} : \overline{KO}$. The temperature then begins to decrease again, but the composition and the amounts of the phases no longer change.

Examples of systems with incongruent melting points are silica-alumina (cristoballite + corundum → mullite), potassium sulphate — cadmium sulphate, potassium chloride — copper (I) chloride, magnesium-nickel and picric acid — benzene.

Type 2-IIa$_1$: the components do not form a compound and are completely miscible in the solid phase (Fig. 3.9).

The phase diagrams of systems that are completely miscible in the solid phase formally resemble liquid-vapour equilibrium diagrams. If the melting points of the mixtures are located between those of the pure components, t_A and t_B, diagrams similar to that given in Fig. 3.9 are obtained. The area of the diagram is divided into three regions: that corresponding to unsaturated solutions, lying above the liquidus line (l), the heterogeneous region located between the liquidus and solidus (s) lines, and the solid solution region located below the solidus line. When the melt represented by point C is cooled, the first crystals of a solid solution with composition F separate at point D. On further cooling, the composition of the liquid phase will change (from point D to the right along the liquidus curve), as well as the composition of the solid phase (from point F to the right along the solidus curve); hence at the temperature corresponding, for example, to point G, crystals of H will be in equilibrium with liquid K in the weight ratio $\overline{GK} : \overline{GH}$.

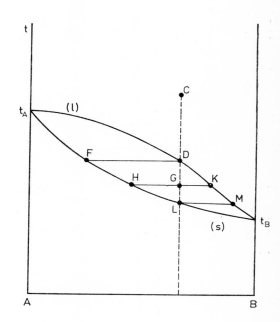

3.9. Phase diagram of 2-IIa$_1$ system type.

The cooling must, of course, proceed so that the change takes place at equilibrium, i.e., the composition of the initially separated crystals, F, changes smoothly up to point H. On further cooling, the point representing the system moves to point L; at this temperature, the last residue of the liquid phase with composition M disappears and the whole system then consists of a solid solution (mixed crystals) with a composition identical with that of the initial system.

With this type of system, pure components can be obtained by fractional crystallization.

Typical examples of systems with complete miscibility in the solid phase are sodium chloride — silver chloride and naphthalene — β-naphthol.

Type 2-IIa$_2$: the components do not form a compound ard are completely miscible in the solid phase (Fig. 3.10).

If the components of a mixture form a series of solid solutions, the melting points of mixtures that lie above or below those of the pure components, t_A and t_B, a diagram with a melting point minimum or maximum is obtained. A diagram with a melting point minimum is depicted in Fig. 3.10.

If a solution characterized by point C is cooled, a solid with composition F begins to separate at point D. On further cooling, the composition of the solid phase being formed changes continuously along the solidus curve between points F and G, the composition of diminishing liquid phase along the liquidus curve between points D and H, until the liquid phase disappears completely, leaving a solid solution (mixed crystals), with a composition identical with that of the initial mixture. By fractional crystallisation, only pure component A and mixture S, corresponding to the melting point minimum, can be obtained. In an analogous manner, when solution K is cooled, first crystals with composition M separate at point L and the compositions of the solid and liquid phases change on further cooling along the solidus curve, MN, and the liquidus curve, LO, respectively. Only pure component B and the mixture with the minimum melting point, S, can be obtained by fractional crystallisation.

A solid solution with the minimum melting point is obtained at point S. This solution constitutes a single homogeneous solid phase and point S therefore differs significantly from the eutectic point.

Typical examples of systems with a melting point minimum are potassium chloride — potassium bromide, arsenic — antimony, gold — silver and potassium nitrate — sodium nitrate. Systems with a melting point maximum occur very rarely.

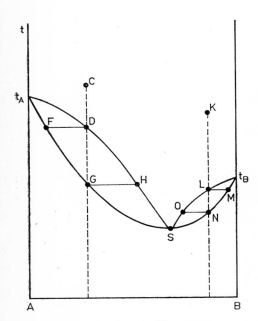

3.10. Phase diagram of 2-IIa$_2$ system type.

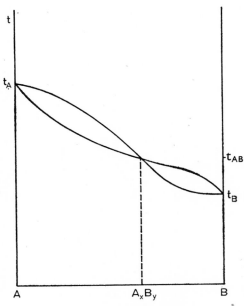

3.11. Phase diagram of 2-IIb system type.

Type 2-IIb: the components are miscible in the solid phase and form a compound (Fig. 3.11).

Phase diagrams of this type can be divided into two diagrams of the 2-IIa$_1$ type for systems A — A_xB_y and A_xB_y — B. Compound A_xB_y and one of pure components A and B, depending on the initial composition of the compound, can be obtained by fractional crystallisation.

Magnesium-cadmium is an example of a 2-IIb system.

Type 2-IIIa: the components are partially miscible in the solid phase (Fig. 3.12).

Systems of this type very often strongly resemble 2-Ia$_1$ type systems, differing from them only in the partial miscibility of the components in the solid phase. If this miscibility is limited to only very low concentrations, it is frequently difficult to distinguish 2-III type systems from 2-Ia$_1$ type systems.

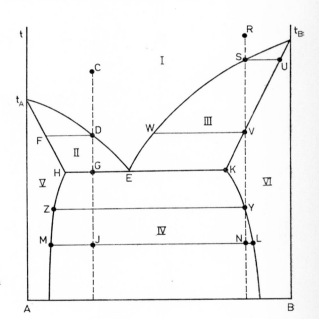

3.12. Phase diagram of 2-IIIa system type.

The areas in Fig. 3.12 denote the following: *I*, the unsaturated solution; *II* and *III*, heterogeneous equilibria between the solid and liquid phases; *IV*, heterogeneous region corresponding to conjugated solid phases; and *V* and *VI*, the heterogeneous solid solutions.

During cooling of the solution characterized by point *C*, a solid phase with composition *F* begins to separate at point *D*. The amount of solid increases on cooling and its composition follows the solidus curve from point *F* to point *H*, while the composition of the disappearing liquid phase varies along the liquidus curve from point D to point E. At the eutectic temperature, corresponding to point E, crystals of H and solution E are in equilibrium. A new solid phase, K, appears at a constant temperature and the cooling is isothermal until liquid E disappears, only the eutectic mixture of crystals H and K remaining, in the weight ratio $\overline{KE} : \overline{HE}$. On further cooling, at a sufficiently slow rate to maintain equilibrium, the composition of the mixed crystals (i.e. the conjugated solid solutions) changes along curves H-Z-M and K-Y-L.

Cooling of solution R proceeds differently. Crystals of solid U begin to separate at point S. During cooling, the amount of solid increases; its composition changes continuously along the solidus curve down to point V and at this temperature the liquid with composition W disappears completely. The whole system contains a single solid — mixed crystals with the original composition of the system. However, on slow cooling down to point Y, the point representing the system reaches the limit of miscibility in the solid and conjugated solid solution Z begins to separate from solid solution Y. The composition of the conjugated solutions again varies with decreasing temperature, so that at a temperature corresponding, for example, to point N, the system will contain a heterogeneous mixture of mixed crystals M and L in the weight ratio $\overline{NL} : \overline{NM}$.

Examples of systems of the 2-IIIa type are nahpthalene — monochloroacetic acid, potassium nitrate — thallium nitrate and iron — chromium.

Type 2-IIIb: Fig. 3.13.

If components A and B in a 2-III type system differ considerably in their melting points, some properties of the phase diagram also change. However, the significance of the areas in Fig. 3.13 is the same as in the previous example.

On cooling solution C, mixed crystals with composition F begin to separate at point D. While the amount of crystals increases, the composition of the liquid changes along the D-K-H curve and that of the crystals along the solidus curve, F-L-G, and finally the whole system solidifies at a temperature corresponding to point G to give a solid solution with the composition of the original system, which no longer changes.

On cooling solution E, mixed crystals L begin to separate at point K, their composition changing with decreasing temperature along curve L-G-U'-N, while the composition of the liquid follows curve K-H-U-P. At point M, when the peritectic temperature has been reached, mixed crystals N are in equilibrium with a solution of composition P. A new solid, O, is formed at constant temperature and its amount increases continuously until the peritectic solution, P, disappears. During further cooling, the compositions of mixed crystals N and O change along the curves of limited miscibility in the solid phase, so that at the temperature of, for example, point Q, a heterogeneous mixture of conjugated solids R and S (homogeneous mixed crystals) results, in the weight ratio $\overline{QS} : \overline{QR}$.

Cooling of solution T proceeds as follows. Solid solution U' begins to separate at point U and while the point representing the liquid moves from point U to point P, the composition of the mixed crystals changes from point U' to point N. At the constant peritectic temperature corresponding to point V, mixed crystals N disappear to produce newly formed crystals of O, so that the system contains an equilibrium mixture of crystals of O and solution P at the end of this cooling

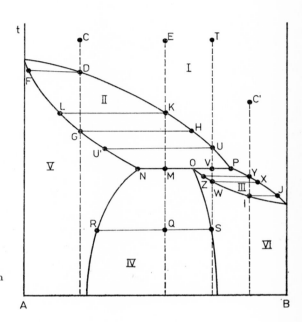

3.13. Phase diagram of 2-IIIb system type.

stage. The composition of the mixed crystals changes on a further decrease in the temperature along the solidus curve, O-Z-W, and the composition of the liquid moves along the liquidus line, P-Y-X, until all of the liquid with composition X disappears at point W. The system then contains only a homogeneous solid solution with composition W and nothing happens on further cooling to point S. At the same temperature as point S, the conjugated solid solution, R, begins to separate from solid solution S and the subsequent changes are the same as in the previous example.

Finally, when solution C' is cooled, solid Z begins to separate at point Y and its composition changes continuously along the solidus line to point I. At this point all of liquid J disappears and the system contains only mixed crystals (a solid solution) with the original composition.

The systems sodium nitrate — silver nitrate and cadmium — mercury are examples of the 2-IIIb type.

3.3. Three-component systems

The appearance of triangular diagrams for a single system may differ, depending on differences in the melting points of the components and on the temperature for which the diagram is valid, i.e., on the place at which the plane representing the temperature intersects the planes representing the phase equilibria.

Three typical situations for the simplest 3-Ia type system are depicted in Fig. 3.14:

(a) solvent C at a temperature above the melting point and two solutes A and B,

(b) solvent C at a temperature below the melting point and two solutes A and B,

(c) three substances with comparable melting points, A, B and C. The significance of the individual areas is identical in the three diagrams:

I, the unsaturated solution region,

II, the region corresponding to the equilibrium of solid A with the saturated solutions;

III, the region corresponding to the equilibrium of solid B with the saturated solutions;

IV, the region corresponding to the equilibrium of solid C with the saturated solutions;

V, the region corresponding to coexistence of crystals of A and B with solution E_1;

VI, the region correspoding to coexistence of crystals of B and C with solution E_2.

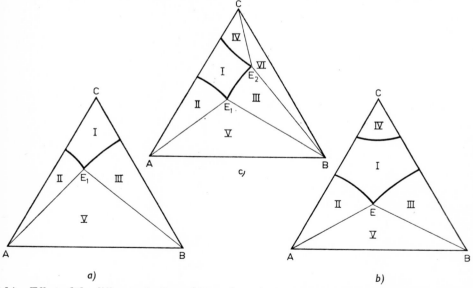

3.14. Effect of the difference in the melting points of substances on the character of the ternary diagram.

When all of the isotherms are depicted up to the eutectic point, the ternary phase diagram assumes the form shown schematically in Fig. 3.15 (the numbers in the diagram denote the individual isotherms).

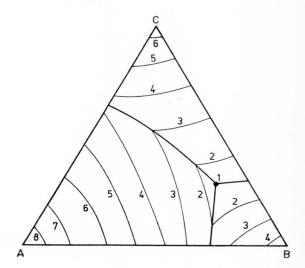

3.15. Isoterms for the whole range of a three-component system.

Type 3-Ia: the components are immiscible in the solid phase and do not form a compound (Fig. 3.16).

The simplest phase diagram of a system consisting of two solutes A and B and solvent R is depicted in Fig. 3.16; for the sake of comparison, the system is represented in equilateral triangular coordinates and then in right-angled triangular coordinates. The significance of the areas is the same as in Fig. 3.14.

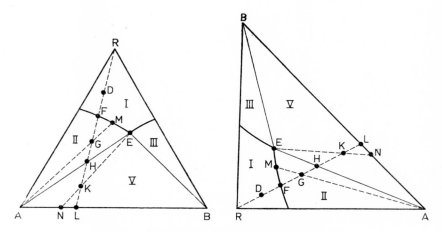

3.16. Phase diagram of 3-Ia system type.

Starting at point D, representing an unsaturated solution, the solvent is evaporated from the solution at a constant temperature. According to the straight-line rule, the composition of the system will thus change so that the point representing the system will move along a line connecting points R and D. At point F, the solubility curve of substance A is reached. On further cooling, crystals of substance A will separate and the composition of the mother liquor will follow curve FME. If the composition of the evaporated solution corresponds, for example, to point G (the ratio of the weight of evaporated water to the weight of the remaining system G equals the ratio of segments $\overline{DG} : \overline{DR}$), the equilibrium composition of the mother liquor lies on the line connecting points A and G, extrapolated to point M. The weight ratio of the crystals to the mother liquor is $\overline{GM} : \overline{AG}$.

Pure substance A separates on evaporation up to point H. At this point, corresponding to the ratio of evaporated water to the original weight of the system, $D = \overline{DH} : \overline{RH}$, the weight ratio of the crystals to the mother liquor is $\overline{HE} : \overline{HA}$ and the composition of the mother liquor is given by the eutonic point, E. The solution is then saturated with components A and B so that crystals of both substances will separate on further evaporation. For example, on evaporation of the solution up to point K, a mixture of crystals with overall composition N is obtained in equilibrium with mother liquor E, in the weight ratio $\overline{KE} : \overline{KN}$. Crystals A and B will be present in this mixture in a weight ratio of $\overline{NB} : \overline{NA}$. Finally, when all of the solvent has evaporated, a mixture of crystals, L, is obtained with an A : B ratio identical with that in the original solution, D.

This type of system is very common. Examples are sodium nitrate — sodium nitrite-water, ammonium chloride — ammonium sulphate — water, trichloroacetic acid — dichloroacetic acid — monochloroacetic acid and potassium chloride — sodium chloride-water.

Type 3-Ib$_1$: the components are immiscible in the solid phase and form a compound (Fig. 3.17).

Two situations can be encountered with this type of system: the solute forms a binary compound with the solvent, e.g. the hydrate A_xR_y (Fig. 3.17, left) or with the other solute, A_xB_y (Fig. 3.17, right). On evaporating solution D, the composition of the solution shifts, analogous to the previous example, along a straight line passing through apex R. Crystals of the binary compound (A_xR_y or A_xB_y) begin to separate at point F; the pure substance separates up to point H, where a mixture of crystals (A_xR_y and B or A_xB_y and B) begins to separate on further evaporation. Evaporation ends at point L.

Examples are sodium sulphate — sodium chloride-water at 15 °C and ammonium nitrate — silver nitrate-water.

An interesting phenomenon can sometimes occur during hydrate formation: the substance crystallises as a hydrate or (at the same temperature) as the anhydrous substance, depending on the content of the other component. An example

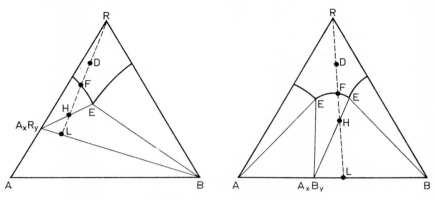

3.17. Phase diagram of 3-Ib$_1$ system type.

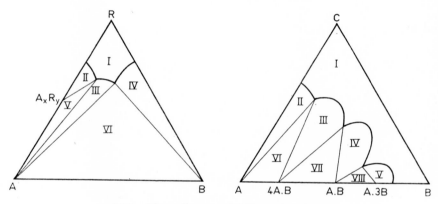

3.18. Phase diagram of 3-Ib$_1$ system type.

of such a diagram is shown on the left-hand side of Fig. 3.18. Region I corresponds to the unsaturated solutions, II to the formation of crystals of hydrate A_xR_y, III to crystallisation of anhydrous salt A, IV to crystallisation of salt B, V to the simultaneous precipitation of A_xR_y and A and VI to the simultaneous precipitation of A and B. The sodium sulphate — sodium chloride-water system at 25 "C is an example of this type of behaviour.

Some systems form a whole series of binary compounds. For example, in the ammonium sulphate-sulphuric acid-water system, the phase diagram for which si depicted schematically on the right-hand side of Fig. 3.18, compounds with component ratios of 4 : 1, 1 : 1 and 1 : 3 are formed. Area I again corresponds to the unsaturated solutions, areas II, III, IV and V to the precipitation of compounds A, 4A.B, A.B and A.3B, respectively, and areas VI, VII and VIII to the coexistence of two salts and the saturated solution. This diagram is also an example of phase equilibrium in a system that consists of components two of them being liquid at the given temperature.

Another type of diagram (Fig. 3.19) represents a system in which substance A and B form a binary compound and substance B also forms a hydrate. When solution D is evaporated, crystals of anhydrous substance A can be obtained in segment $\overline{MH}$ and crystals of A and A_xB_y in segment $\overline{HS}$; when solution F is

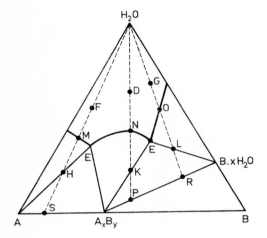

3.19. Phase diagrams of 3-Ib$_1$ system type.

evaporated, crystals of binary compound A_xB_y are obtained in segment $\overline{NK}$ and a mixture of A_xB_y and B . xH$_2$O in segment $\overline{KP}$. Finally, evaporation of solution G leads to crystallisation of B . xH$_2$O in segment $\overline{OL}$ and of a mixture of A_xB_y and B . xH$_2$O in segment $\overline{LR}$.

Type 3-Ib$_2$: the components are immiscible in the solid phase and form a compound (Fig. 3.20).

Fig. 3.20 shows a three-component system in which two components form a binary compound (such as a hydrate) and all three components form a ternary compound, $A_xB_yC_z$, which is depicted as point D in the diagram. The individual areas represent: I, unsaturated solutions; II crystallisation of substance A; III, crystallisation of ternary compound $A_xB_yC_z$; IV, crystallisation of binary compound B_xC_y; V, simultaneous crystallisation of A and $A_xB_yC_z$; VI, simultaneous crystallisation of $A_xB_yC_z$ and B_xC_y; VII, coexistence of solid mixture $A_xB_yC_z$, B_xC_y and B; and VIII, coexistence of solid mixtures $A_xB_yC_z$, A and B.

Examples of this type of system are substances that form alums, such as potassium sulphate-aluminium sulphate-water.

Type 3-IIa: the components are completely miscible in the solid phase (Fig. 3.21).

The diagram is separated into two parts by the solubility curve. The part near apex C represents unsaturated solutions, while that below the curve corresponds to the coexistence of the solid and liquid phases and the base of the triangle,

AB, represents the solid phase. The equilibrium compositions of the solid and liquid phases are interconnected by straight lines (conodes).

If solvent C is evaporated from system S, a solid phase with composition given by point G begins to separate at point F. On further evaporation, the composition

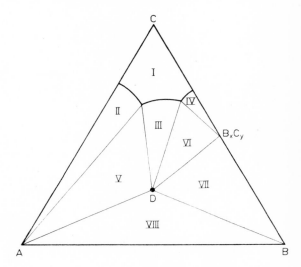

3.20. Phase diagram of 3-Ib$_2$ system type.

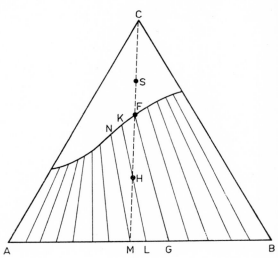

3.21. Phase diagram of 3-IIa system type.

of the liquid moves towards point K and the equilibrium composition of the solid phase simultaneously changes continuously, so that a system represented, for example, by point H, contains solution K and crystals L in the weight ratio $\overline{HL} : \overline{AK}$. After complete evaporation of solvent C, all of liquid phase N disappears and only crystals (solid solution) M remain.

A typical example of this type of system is potassium terephthalate — ammonium terephthalate — water and a number of other systems containing ions that are readily interchangeable.

Phase equilibria in systems that form solid solutions are often represented in the form of x-y diagrams: the content of one component in the solid phase is plotted on the abscissa and that of the same component in the residue after evaporation of the equilibrium solution is plotted on the ordinate. Four typical examples are depicted schematically in Fig. 3.22:

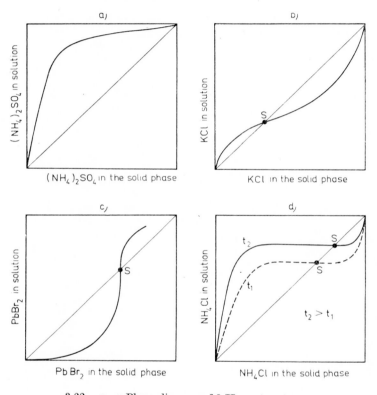

3.22. $x-y$ Phase diagram of 3-IIa system type.

(a) The ammonium sulphate-potassium sulphate-water system, which is characterized by a positive deviation throughout the whole concentration range; thus the crystals contain relatively less ammonium sulphate than the solution.

(b) The potassium chloride-potassium bromide-water system with the shape of a sigmoid curve. At point S, the solid solution apparently behaves as a compound, i.e., the two components are present in the same ratio in the solution as in the solid.

(c) The lead (II) bromide — lead (II) chloride-water system, which also has a sigmoid curve, but reversed compared with the previous case.

(d) The ammonium chloride-potassium chloride-water system, which behaves in an analogous manner to system (b). Here two isotherms are constructed, so that it is evident that the composition corresponding to point S in system (b) changes with the temperature.

Type 3-IIb: the components are partially miscible in the solid phase (Fig. 3.23).

The significance of the individual areas is as follows: I, unsaturated solutions; II, separation of solid solutions of A in B; III, separation of solid solutions of B in A; and IV, corresponds to the coexistence of solid solutions K and L and liquid E.

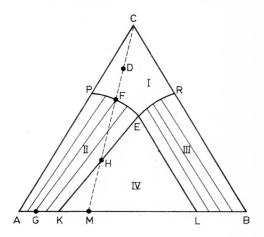

3.23. Phase diagram of 3-IIb system type.

If solvent C is evaporated from solution D, the first mixed crystals with composition G begin to separate at point F. While the composition of the solution changes from point F *to* point E on further evaporation, the equilibrium composition of the mixed crystals changes from point G to point K. At point H, solution E and mixed crystals K are in equilibrium. During further evaporation of solvent C, another solid phase, L, appears in the system, the composition of solution E remaining unchanged. Finally, after evaporating all of the solvent, a heterogeneous mixture of homogenous crystals of solid solution K and L is obtained, in the weight ratio $\overline{LM} : \overline{KM}$.

The 3-IIb and 3-Ia type of phase diagrams are analogous to a certain extent.

3.4. Four-component systems

Type 4-I: four different components (Fig. 3.24).

Fig. 3.24 depicts schematically the phase diagram of the somewhat complicated system, sodium sulphate — potassium sulphate — magnesium sulphate-water at 25 °C. A number of hydrates and binary and ternary compounds are formed in this system: firstly, sodium sulphate exists as the decahydrate and magnesium sulphate as the heptahydrate; in the ternary system sodium sulphate-magnesium sulphate-water, the ternary compound astrachanite, $Na_2SO_4 \cdot MgSO_4 \cdot 7H_2O$ (point I) is formed; in the ternary system magnesium sulphate-potassium sulphate-water, the ternary compound schönite, $K_2SO_4 \cdot MgSO_4 \cdot 6H_2O$ (point II) is formed; finally, the binary compound glazerite, $Na_2SO_4 \cdot 3K_2SO_4$ (point III) is formed in the ternary system sodium sulphate-potassium sulphate-water.

According to the Gibbs phase law eqn. (1.60), for an isothermal four-component system

$$v + f = 4 \tag{3.1}$$

and a solution saturated with three salts will be represented by a single point. Four such points can be found in Fig. 3.24: B, D, G and E. In addition, points C,

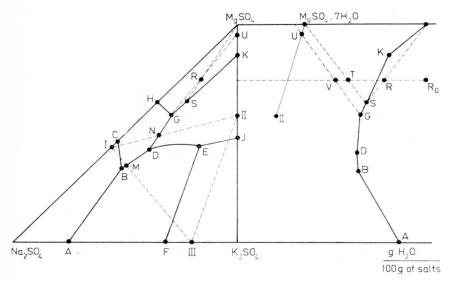

3.24. Phase diagram of 4-I system type.

H, K, J, F and A are given in the graph for solutions saturated with two salts in ternary systems. This system of points is connected by curves, dividing the overall phase diagram into a number of regions:

Sodium sulphate — $A — B — C$:   corresponds to $Na_2SO_4 \cdot 10H_2O$.
$A — B — D — E — F$:   corresponds to glazerite as an equilibrium solid phase.
Potassium sulphate — $J — E — F$: corresponds to K_2SO_4.
$J — E — D — G — K$:   corresponds to schönite as an equilibrium solid phase.

$C - B - D - G - H$: corresponds to astrachanite as an equilibrium solid phase.

Magnesium sulphate — $K - G - H$: corresponds to $MgSO_4 \cdot 7\,H_2O$.

Isothermal evaporation of any initial solution must end in the point representing a solution saturated with the three substances, on the condition, of course, that this point is congruent, i.e., that it represents a real phase equilibrium at the given temperature. Such a point must be located in a space diagram in the space limited by planes connecting possible equilibrium phases and the same must hold for any projection of the appropriate isotherm.

The dashed lines connecting points I, II and III in Fig. 3.24 divide the whole phase diagram into four triangles. Point B, representing a solution saturated with sodium sulphate, I and III, lies inside the triangle determined by these three points, i.e., is congruent. Similarly, point D lies inside triangle $I - II - III$ and hence is also congruent. As point G lies in triangle $I - II$ — magnesium sulphate, it is also congruent. The fourth invariant point is E, representing a solution saturated with II, III and potassium sulphate; however, it is located outside region $II - III$ — potassium sulphate and hence is incongruent, and therefore isothermal evaporation of the four-component system cannot end there.

Evaporation of solutions whose representative points lie on segments $\overline{GK}$ and $\overline{HG}$ leads unambiguously to point G. Systems represented by points on segments $\overline{AB}$ and $\overline{BC}$ will proceed to point B on further evaporation. Systems represented by points on segments $\overline{JE}$ and $\overline{FE}$ will proceed to point E on evaporation; however, as this is an incongruent point, potassium sulphate will gradually dissolve and the point representing the system will move from point E to the final point, D.

At the intercepts of straight lines $\overline{BD}$ and $\overline{DG}$ with the dashed lines connecting the points corresponding to compounds I, II and III, characteristic congruent points M and N are found. If the composition of the solution evaporated is given by one of the two points, that ratio of the components in the liquid and solid phases will be constant during the whole evaporation. If the composition of the solution is represented by a point on segment $\overline{BD}$, other than point M, the point representing the system will move away from point M during the evaporation. The same holds for segment $\overline{DG}$ and point N.

Finally, isothermal evaporation of water from the system corresponding to point R can be represented. Point R lies inside the region corresponding to separation of $MgSO_4 \cdot 7\,H_2O$ and can thus be connected with the apex $MgSO_4$. This line intersects the curve corresponding to solutions saturated with two salts at point S corresponding to a solution saturated with II and magnesium sulphate simultaneously. In the "aqueous" projection on the right-hand side it can be seen that original solution R_0 was unsaturated; after evaporation of amount of water corresponding to segment $\overline{R_0R}$, the solution becomes saturated and $MgSO_4 \cdot 7\,H_2O$ begins to crystallise out. On further evaporation, the amount of water evaporated corresponding to $\overline{RT}$, crystals of $MgSO_4 \cdot 7 H_2O$ continue to precipitate until the composition of the mother liquor arrives at point S. The weight ratio of the crystals to the mother liquor is given by the $\overline{SR} : \overline{R\text{-}MgSO_4}$ ratio. On further evaporation, $MgSO_4 \cdot 7 H_2O$ and II will precipitate simultaneously and the composition of the mother liquor will move to point G. This liquid phase composition corresponds to the composition of an equilibrium mixture of two salts ($MgSO_4 \cdot 7 H_2O +$ II), represented by point U. The amount of water evaporated from the system during this stage corresponds to segment $\overline{TV}$ in the aqueous projection. It is now possible to evaporate further portions of water at point G at a constant mother liquor composition and three solids will crystallise simultaneously: $MgSO_4 \cdot 7 H_2O$, II and I. The overall solid phase composition will shift from point U (at the beginning of the evaporation) to point R (complete evaporation of water).

Type 4-II: reciprocal salt pair (Fig. 3.25).

A reciprocal salt pair system

$$AX + BY \rightleftarrows AY + BX \qquad (3.2)$$

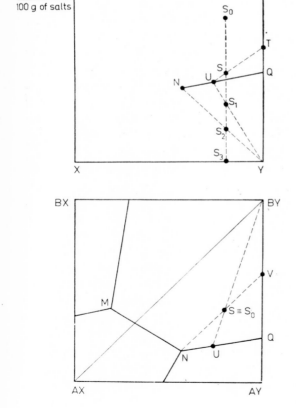

3.25. Phase diagram of 4-II system type.

is represented in Fig. 3.25 at the temperature at which both invariant points, M and N, are located in the appropriate triangles, AX-BY-BX and AX-BY-AY, respectively, in other words are congruent. Isothermal evaporation of a solution with composition S_0 can be represented in this diagram. As can be seen in the "aqueous" projection in the upper part of the figure, solution S_0 is unsaturated. In this projection point T represents a solution that is just saturated with pure salt BY, i.e., with the salt in whose separation region point S lies. The line connecting point $S_0 \equiv S$ with apex BY intersects the solubility curve at point U and if points U and T in the aqueous projection are connected, point S, which represents a solution saturated with substance BY, is obtained. The amount of water evaporated is then proportional to segment $\overline{S_0 S}$. When a further portion of water is evaporated, crystals BY separate and the point representing the solution shifts to point U. At this point the solution is saturated with salts BY and AY. The amount of water evaporated in this stage corresponds to segment $\overline{SS_1}$. On further evaporation, crystals of two solids, AY and BY, will separate simultaneously and the composition of the solution will vary along segment $\overline{UN}$. After evaporation of water corresponding to segment $\overline{S_1 S_2}$, the point representing the solution shifts to point N (a solution saturated with three components, BY, AY and AX), and the point representing the sum of the solid phases moves to point V. The amount of solution and solid phase are in the ratio $\overline{SV} : \overline{NS}$ and salts BY and AY in the solid phase are in the ratio of segments $\overline{V\text{-}(AY)} : \overline{V\text{-}(BY)}$. During evaporation of water in the interval $\overline{S_2 S_3}$, crystals of AY, AX and BY separate at constant solution composition, N.

Fig. 3.26 depicts two isotherms for a system of the 3-II type, where t_1 (solid line) $< t_2$ (dashed line). Isotherm t_1 is identical with the isotherm given in Fig. 3.25; isotherm t_2 differs chiefly in the fact that one of the invariant points, R, is not located inside the corresponding triangle, AX-BY-BX, and is thus incongruent. The cooling of two typical solutions can then be represented in this diagram.

If a solution saturated with substances AY and BY at a higher temperature t_2, represented by point G, is cooled down to temperature t_1, then the point representing the system will be located in region D-M-N-O-BY, i.e. in the region corresponding to the separation of salt BY. The line connecting point BY with point G intersects the corresponding phase equilibrium curve at point H. Of course, solution G was saturated with both salts, AY and BY; on cooling, both salts will separate, corresponding to point Y in the "aqueous" projection. The line connecting Y with point G intersects isotherm $\overline{DN}$ at point K, which is not identical with point H (if the two points were identical, only substance BY would separate). Point K in the "anhydrous" projection represents the real composition of the mother liquor and the line connecting it with point G determines the total composition of solid L, which is formed by crystals of AY and BY in the weight ratio $\overline{L\text{-}(BY)} : \overline{L\text{-}(AY)}$. If only pure salt BY were to be obtained from the solution by crystallization, the initial solution would have to be diluted so that points H and K

would merge. The amount of water that would have to be added to original solution G would correspond to segment $\overline{GF}$, where point F lies on the line connecting points H and Y.

In the second example, S is the starting point; it represents a solution saturated with salts BY and AX at temperature t_2. If this solution were evaporated at constant temperature t_2, salts AX and BY would separate and the composition of the solution would change continuously up to point P, as point R is incongruent. At point P, the third salt, AY, would begin to precipitate in addition to the other two salts. If solution S is cooled down to temperature t_1, either a mixture of salts BY and AX or salt BY alone will crystallise out depending on the amount of water in the original system. If separation of pure salt BY from the solution were required, the original solution, S, would have to be diluted with water in an amount corresponding to segment $\overline{SS_1}$.

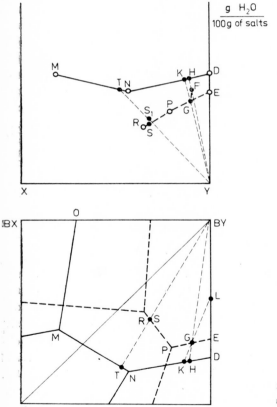

3.26. Cooling process in the 4-II system.

4. MEASUREMENT OF PHASE EQUILIBRIA IN CONDENSED SYSTEMS

Methods of measuring phase equilibria in condensed systems can be classified as static or dynamic. In static methods, it is assumed that phase equilibrium is actually established in the heterogeneous system. According to the procedure employed, they can be subdivided into analytical and synthetic methods. Dynamic methods follow the behaviour of the system during continuous variation of the determining parameters while a change in the behaviour of the system occurs around the phase equilibrium. These methods do not assume establishment of phase equilibria and time appears as one of the monitored variables.

It is impossible to recommend a single method that is universal for any possible type of system. For each system, it is necessary to choose, on the basis of various considerations and sometimes of preliminary experiments, that method which yields the most reliable results in a given instance.

4.1. Analytical Methods

The analytical methods that can be employed are very simple in principle. The components are mixed in an approximate ratio so that one solid phase is in excess at the required temperature and the system is closed and maintained thermostattically at constant temperature for a sufficient time to ensure establishment of equilibrium; then samples of the solid and liquid phases are taken in a suitable manner and analyzed. However, several problems can be encountered during the practical performance of the experiments that complicate the utilization of the method. Basically, these problems are the following:

(a) weighing the components in the correct ratio, so that the solution is in equilibrium with a single solid phase;

(b) designing an apparatus that permits reliable establishment of the equilibrium state;

(c) determination of the time required for establishment of equilibrium;

(d) sampling of the equilibrium phases;

(e) determination of the composition of the equilibrium phases (including a correction for mother liquor adhering on a solid).

These problems are discussed in greater detail below.

(a) The components should be added in a suitable ratio so that amounts of both phases sufficient for sampling are obtained. It is clear that the solid phase must be present in excess at the temperature of measurement. However, this must be a single solid phase if a point on the solubility curve is to be determined. If two solid phases are present in excess in the system, the composition of the eutonic point solution is found by analysis of this solution; analysis of the solid phase gives no results that can be used for the construction of a phase diagram. If the amount of solid phase is too large, difficulties are encountered during sampling of the liquid, and vice versa. In addition, the rate of establishment of phase equilibrium is proportional to the area of the interface and can therefore be very slow with too small an excess of the solid phase. For the same reason, the individual solid components are weighed in the form of a very fine powder, as small particles have a large specific surface area.

(b) During measurement of phase equilibria by an analytical method, it is necessary to ensure that the system remains at a constant temperature, that there is good contact between the phases in equilibrium and that there is a uniform concentration distribution in the bulk of the solution. The required precision of temperature maintenance depends on the properties of the system being measured, (the temperature dependence of the solubility) and should usually not be worse than $\pm$ 0.1 °C. Stirring the system with a stirrer is not recommended, as it is difficult to prevent evaporation of the more volatile components. Stirring with an air bubble in a sealed or tightly stoppered vessel, which is rocked or rotated along its horizontal axis in a thermostat, is most frequently used. Two satisfactory means of fixing the samples in the thermostat are depicted schematically in Fig. 4.1. In the upper part are depicted bottles that contain weighed mixtures, closed with rubber stoppers and fixed with screws in a turnable frame immersed in a water thermostat. At the bottom is depicted the fixing of sealed test-tubes with the aid of steel or bronze springs to an axle immersed at an angle in a water thermostat.

(c) Establishment of the equilibrium takes very different lengths of time, depending on the type of system involved. This process can take a few tens of minutes but, on the other hand, instances have been described in the literature (e.g. determination of the solubility of calcium sulphate) in which weeks or even months were required. It is generally recommended that the solid phase should be weighed in a finely crystalline form and that the system should be stirred regularly during the whole equilibration time. Vibrations, ultrasonics and similar effects hasten the process. Sometimes, especially when recrystallisation of a solid phase takes place simultaneously, the establishment of equilibrium can be hastened by moderate variations in the temperature. With substances whose solubility increases with increasing temperature, it is recommended that the equilibrium should be approached from a lower temperature (heating to the required temperature); with the reverse procedure, a supersaturated solution could be formed and the supersaturation is sometimes eliminated very slowly.

The time required for establishment of equilibrium is determined in a series of preliminary experiments; samples of the phases are taken at various intervals and the time after which their composition no longer changes is found. In the measurement itself, a sufficient safety margin is added to this time (100 % or more).

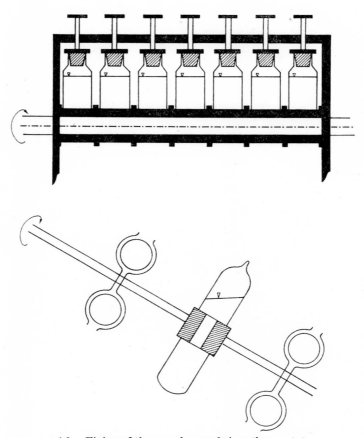

4.1. Fixing of the sample vessels in a thermostat.

d) The separation and sampling of the equilibrium phases are usually performed by using a device such as those shown in Fig. 4.2. The liquid phase is sampled with glass filters (a) or cotton wool filters (b, d, e). In exceptional instances, when the solid settles easily, the transparent liquid is carefully decanted into another vessel (f). The solid phase is usually sampled with glass filters (c, g). In order not to disturb the equilibrium, either phase separation is carried out in the thermostat or the sampling device is controlled thermostattically at the phase equilibrium temperature. The solid phase always contains adhering mother liquor, for which a correction must be made. In some correction procedures, care must be exerted that no change in the overall composition occurs due to evaporation of the more volatile component (the solvent); therefore, the sample is separated very quickly by filtration at decreased or, preferably, increased pressure (or by centrifugation) and then transfered carefully into a closed weighing bottle. Unnecessary evapora-

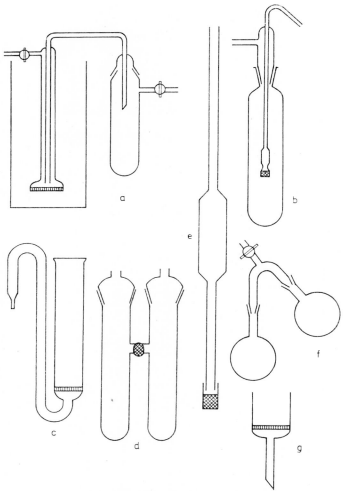

4.2. Devices for separation of equilibrium phases.

tion of the solvent from the liquid phase should also be avoided and therefore the sample is placed in, for example, a weighing bottle and weighed immediately.

(e) As already mentioned, the solid phase always contains adhering mother liquor and therefore a correction must be made when determining its true composition. The correction for adhering mother liquor can be performed by using one of the following four methods: (i) The damp solid phase is weighed (g_1) and .

then dried to a constant weight (g_2). The drying must be carried out carefully in order to prevent the decomposition of any hydrate that might be present. If the mother liquor contains (according to liquid phase analysis) p_0 % of solvent, then the amount of adhering mother liquor is

$$g_r = 100(g_1 - g_2)/p_0. \tag{4.1}$$

From the analysis and the known amounts of the damp solid and the adhering mother liquor and using a mass balance, the actual composition of the equilibrium phase can then easily be calculated.

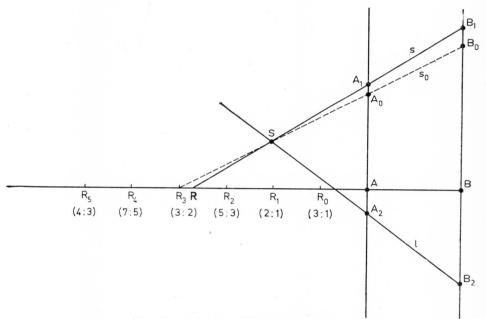

4.3. Correction for adhering mother liquor.

(ii) A small amount of an inert substance, which is virtually without effect on the equilibrium and which is not transferred into the solid phase, is added to the system. Assuming that the added substance is not adsorbed to a great extent on the surface of the crystals, we can determine the amount of mother liquor adhering on the crystals from the contents of the substance in the liquid and in the damp solid phase, and can carry out the appropriate corrections to the composition of the damp solid phase. (iii) The third method depends on the fact that the components are frequently contained in the solid phase in a ratio corresponding to small integers. The crystals are quickly filtered off under vacuum so that they contain less than 10 % of adhering mother liquor and the liquid and damp solid are analyzed. The composition of the pure solid phase is found graphically according to Fig. 4.3. The number of moles of substances A and B in the damp residue are plotted vertically upwards from points A and B, respectively, thus obtaining points A_1 and B_1. From these points, the numbers of moles of substances A and B in the solution (or an arbitrary multiple thereof) are plotted downwards, thus obtaining points A_2 and B_2, respectively. The intercept of straight lines s and l is denoted as point S. Straight line s intersects the abscissa at point R, for which it holds that

$$\overline{AA_1} : \overline{BB_1} = \overline{RA} : \overline{RB}, \tag{4.2}$$

A number of points, $R_0, R_1, R_2, \ldots$, are plotted on the abscissa to the left of point A that give the ratio of the components, B : A, in the compound, corresponding to a ratio of small integers, e.g., 3 : 1, 2 : 1, etc., according to the equation

$$\overline{R_iA} = \frac{\overline{AB}}{(B/A)_i - 1}. \tag{4.3}$$

The point closest to the intercept of straight line s with the abscissa is connected with point S. Thus straight line s_0 is obtained and can be used to correct the composition of the solid phase for adhering mother liquor, (A_0, B_0). In binary systems, it is assumed that B is the solvent.

This method has the disadvantage that the point corresponding to a 2 : 1 ratio is far from point A and the point corresponding to a 1 : 1 ratio is at infinity.

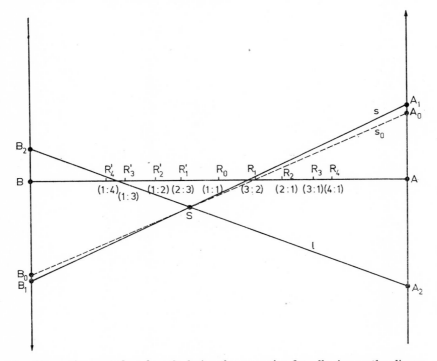

4.4. Alternative procedure for calculating the correction for adhering mother liquor.

The intercept of straight line a with the abscissa is determined with very poor precision for these small ratios. For this reason, a modification of this method, represented in Fig. 4.4, is preferable. The points are denoted in the same manner as in the previous example; the difference in the construction is that the B-axis is drawn in the opposite direction to the A-axis. Points R_i and R_i' divide the straight line $\overline{AB}$ in the ratio of the components in the solid phase so that

$$B : A = \overline{R_iB} : \overline{R_iA}. \tag{4.4}$$

An analogous procedure can be used when a stoichiometric binary compound if formed in a three-component or multicomponent system.

(iv) In three-component and multicomponent systems, Schreinemaker's method of wet residues is most frequently employed. The method is based on the application of the straight line rule: points characterizing the liquid phase, damp solid phase and the pure equilibrium solid phase must lie on a single straight line. If we construct straight lines for two different initial systems connecting points that represent the liquid phase and the wet residue and if identical equilibrium solid phases

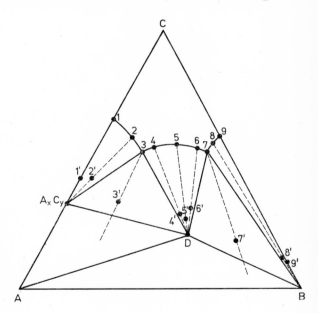

4.5. Principle of Schreinemaker's method.

are formed in the two systems, then the two straight lines intersect at the point corresponding to the equilibrium solid phase. The application of Schreinemaker's method is shown in Fig. 4.5. Open circles denote the composition of the liquid phase; full circles correspond to the composition of the wet residue. Straight lines 1 and 2 intersect at point A_xC_y, straight lines $4, 5$ and 6 at point $D = A_xB_yC_z$ and straight lines 8 and 9 at point B. Solutions 3 and 7 correspond to eutonic points, since the composition of solid phases $3'$ and $7'$ does not correspond to an intercept of the other straight lines. When using Schreinemeker's method, points sufficiently far apart must be chosen in order to locate their intercept as accurately as possible. Straight lines that are too close (e.g. *8* and *9*) intersect at too small an angle so that even a small analytical error can cause a large error in the determination of the intercept. On the other hand, points too far apart should not be chosen, as such systems need not have a common equilibrium solid phase (e.g. points *2* and *4*). Finally, a third disadvantage of Schreinemaker's method is that evaporation of the solvent, C, from the samples of the solid phase must be prevented in order to avoid an undesirable shift in the points that represent the wet solid phase and thus also a change in the slopes of the corresponding straight lines.

The analytical method is disadvantageous in that it is tedious and time consuming; its advantages lie in the possibility of measuring a large number of samples simultaneously and in that the method yields reliable data, even on solid phase composition.

4.2. Synthetic methods

Knowledge of the equilibrium composition of the liquid phase sometimes suffices for construction of the phase diagram. The composition of solid phases that can occur in the given system is frequently known and, in addition, the shape of the solubility curve generally indicates the possible occurrence of any other solid phase. Of course, the components of the system must not form mixed crystals (solid solutions). Data on the composition of the equilibrium liquid phase could be obtained by a standard analytical procedure, taking samples from the liquid phase alone, but these data can be obtained much faster and more easily by means of a synthetic method.

Synthetic methods are based on weighing or measuring the individual components to obtain a system with a known composition; the state in which the solid phase just disappears is then determined for this system. Disappearance of the solid phase can be achieved either by a change in the temperature (polythermal methods) or by the addition of a known amount of solvent (isothermal methods). The disappearance of the solid phase can be monitored either visually or using the physico-chemical and physical properties of the system.

4.2.1. *Polythermal methods*

The components of the system are weighed into a glass vessel (test tube, flask) in a ratio corresponding roughly to the composition of a saturated solution in the required temperature range, so that at least one solid phase is present in excess at the lowest temperature. The vessel is closed (sealed or stoppered), placed in a thermostat and the temperature is slowly increased, the sample being stirred constantly. A visual check is occasionally made so as to ascertain whether the solid phase is still present. When only a few crystals remain, the temperature is maintained constant for sufficient time to ensure that these crystals do not dissolve. Then the temperature is increased minutely and the process is repeated until the exact temperature at which the last of the solid phase dissolves is determined. The same procedure is followed for a number of systems with different compositions. The points that represent the systems are plotted on a graph and the temperatures found are supplemented as coordinates. By interpolation (perpendicular to the expected shape of the isotherm, if possible), a number of points corresponding to the given isotherm are obtained. The interpolation procedure is shown in Fig. 4.6. The full circles with coordinates denote the experimental data, the dashed straight lines join the points between which the interpolation was carried out, and the crosses denote the interpolated values with a coordinate of 40 °C.

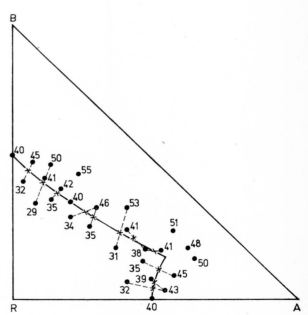

4.6. Interpolation procedure in the polythermal method.

In less accurate measurements, the equilibrium temperature can be determined by slowly heating the heterogenous system placed, for example. in an erlenmeyer flask and stirred with a magnetic stirrer.

The instant at which the crystals dissolve can, of course, be determined not only visually but also by a number of physico-chemical methods. During crystals on, the concentration of the solution increases so that all of those properties of the solution which are dependent on the solute concentration change continuously (e.g., the conductivity, refractive index, density and vapour pressure); after dissolution of the last crystal, only the temperature dependence of the value monitored is observed so that the character of the dependence suddenly changes. For example, Fig. 4.7 shows the change in the refractive index during

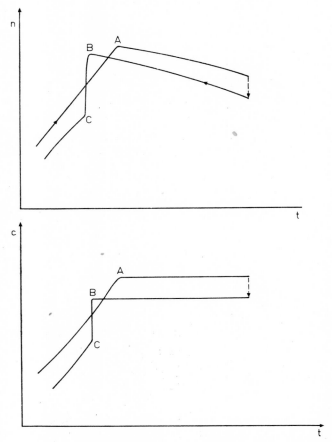

4.7. Change in concentration and refractive index during heating of a suspension and cooling of the corresponding solution.

heating of the suspension (upper curve) and cooling of the solution (lower curve), recorded with a recording refractometer (Czechoslovak Patent 124 740). Fig. 4.7. also shows that precipitation of the crystals takes place at a lower temperature than their dissolution. During heating of the suspension, the crystals are dissolved, the concentration of the solution increases and thus the refractive index increases relatively steeply. As soon as the last crystals dissolve (point A), the concentration of the solution remains constant and the refractive index starts to decrease slowly with increasing temperature. During cooling of the same solution (the refractive index curve is shifted downwards for the sake of clarity), the refractive index first increases slightly, up to point B. Here a large number of crystals separate, the concentration of the solution decreases sharply, as does the refractive index (up to point C), and on further cooling the solid phase separates according to the solubility curve.

Comparison of the two curves reveals an important characteristic of the polythermal method. During slow heating it is impossible to superheat the suspension, i.e., the crystals are dissolved at the equilibrium temperature; however, during cooling of the solution, supercooling frequently occurs and therefore the crystals appear at a temperature lower than the equilibrium temperature. Measurement of phase equilibria with substances whose solubility increases with increasing temperature must always be carried out with increasing temperature. On the other hand, with substances whose solubility decreases with increasing temperature, the measurement must be carried out with decreasing temperature. Thus the equilibrium in the polythermal method must always be determined by monitoring the disappearance of one of the phases of the heterogeneous system.

The Töppler method is based on a different principle. If a crystal is immersed in a solution just saturated with the same substance, the solution in the vicinity of the crystal is optically homogeneous. If the solution is not saturated, the crystal starts to dissolve and a concentration gradient is formed in its vicinity; the concentration will be highest immediately adjacent to the crystal surface and will decrease in the direction of the solution. Thus a thin layer at the crystal surface will be optically inhomogeneous. Similarly, in a supersaturated solution, the crystal will grow and a concentration gradient in the opposite direction will be formed; the layer of solution at the surface of the crystal will again be optically inhomogeneous. This optical inhomogeneity can easily be monitored by using the apparatus shown in the upper part of Fig. 4.8. The instrument consists of a point-source of light, Z, placed at the focus of a collimator, L_1. Lens L_1 focuses the light rays into a parallel beam, which passes through a thermostattically controlled cuvette containing the test solution in which a crystal, K, a thermometer, T, and a stirrer, M are immersed. The light is then concentrated by a projection lens, L_2, placed so that the measuring crystal is sharply projected on to a screen, S. At the focus of the lens is placed shutter C, which obscures half of the beam. The beam path for unsaturated and supersaturated solutions is shown in the lower part of Fig. 4.8. It can be seen that, owing to optical inhomogeneity in solutions that are not completely saturated, a light strip appears on the image of the crystal on the other side of the shutter; the strip disappears just as the solution becomes saturated. In the literature[1] it is stated that the equilibrium temperature can be determined with a precision of better than 0.1 °C; in our experience the error is usually substantially higher.

[1] J. Mýl and J. Kvapil, Collect. Czech. Chem. Commun. 25, (1960) 194.

A modification of the Töppler method has been described in the literature.[2] The thermostattically controlled solution is pumped into a Plexiglass chamber (Fig. 4.9), where an ingenious system ensures complete homogeneity of the tem-

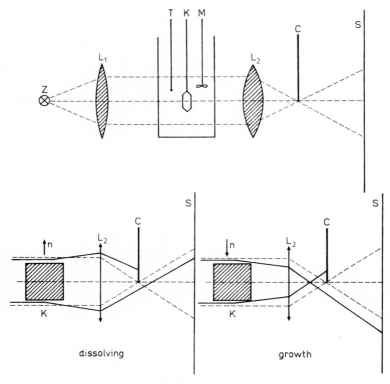

4.8. Töppler method.

perature and concentration around the measured crystal. The temperature of the solution can be varied slowly and continuously, as in the previous instance. The crystal is irradiated with a coherent light beam from a slanted slit and is observed along the axis of the light beam. According to the state of the solution, a light line, broken at a wide or small angle with respect to the split, appears on the vertical face of the crystal; the line is invisible in a solution that is just saturated. According to the data given by the authors, it is possible to determine the equilibrium temperature with a precision of $\pm 0.1\ °C$ for substances with solubilities of about 50 wt. — %; with substances whose solubility ist about 10 wt. — %, the precision decreases to $\pm 0.5\ °C$. These data also represent the upper attainable precision limit and can be considered as realistic. The authors

[2] L. A. Dauncey and E. Still, J. Appl. Chem., 2 (1952) 399.

state that, instead of a crystal, a pellet obtained from the powdered substance by compression with a piece of damp paper pressed on the measuring plane for a short period before the measurement can be used.

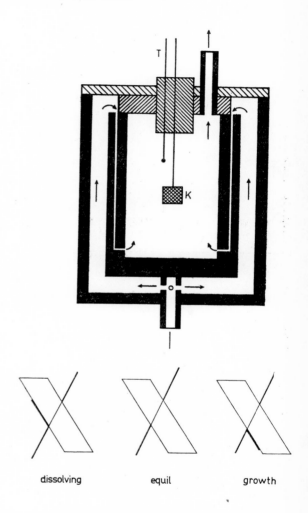

4.9. Modification of the Töppler method.

dissolving   equil   growth

With both of the modifications of the Töppler method discussed, it is unimportant from a theoretical point of view whether the measurement is performed with increasing or decreasing temperature (in contrast to the other polythermal methods). However, much of the data published in the literature was obtained by a common method during a temperature decrease (or even by a thermometric method, a break in the time dependence of the decreasing temperature indicating crystal separation) and therefore such data must be treated with reserve and checked carefully.

4.2.2. Isothermal methods

As already mentioned, the disappearance of the solid phase can also be achieved by gradual addition of the solvent at a constant temperature. The solvent is added in small increments to a termostattically controlled heterogeneous system

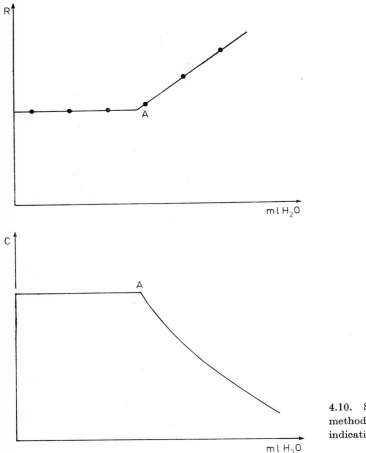

4.10. Synthetic isothermal method with conductivity indication.

with known composition (prepared by weighing the components); after each addition the system is stirred for a long period at a constant temperature and the procedure is repeated until the last of the solid phase disappears. The equilibrium composition of the liquid phase is calculated from the initial component content and the amount of solvent added.

As in the previous instance, the disappearance of the solid phase can be monitored both visually and by measuring various physico-chemical quantities. In the latter instance, the solvent can be added in regular increments and the amount corresponding to the disappearance of the solid phase can be found by extrapolation of the linear sections of the dependence measured. For example, the shape of a conductometric signal is represented in Fig. 4.10; the resistance, R, of the solution does not change unless the concentration changes, i.e., solvent addition is compensated for by addition of solid phase. After the disappearance of the crystals, the solution is diluted by the addition of the solvent and the measured resistance increases. The intercept, A, of the straight lines passing through the experimental points corresponds to the disappearance of the last of the solid phase. The shape of the curve can be different with multicomponent systems, but the disappearance of the solid phase is always indicated by a break.

The method described is simple, rapid and yields data for construction of the equilibrium isotherm. It has, of course, the disadvantage common to all synthetic methods, i.e., it does not yield information on the equilibrium solid phase or phases directly. In this method, it is also necessary to add the solvent sufficiently slowly in order that equilibrium can be established; the acceptable rate of addition of solvent must be determined either by a series of preliminary experiments or (e.g. in conductivity measurements) additional solvent is added only after stabilization of a constant value of the measured quantity.

4.3. Dynamic methods

While in the previous methods the temperature or composition dependence of either the phase equilibrium itself or of the steady-state behaviour of the system was followed, dynamic methods are based on monitoring the properties of a system in their dependence upon a deviation from equilibrium. Time then appears as a new independent variable in the measurement. In this way, it would obviously be possible to monitor the time dependence of any property of a system brought out of equilibrium. In this section, however, only two basic methods will be discussed in detail.

4.3.1. Cooling curve

In Chapter 3, the behaviour of various types of systems during cooling was discussed and we pointed out that the separation of a substance proceeds at a constant temperature at certain characteristic points in the diagram, while the heat removed by cooling is compensated for by the latent heat of the phase change. However, it can be expected that cooling will generally proceed (at a constant rate of heat removal) at different speeds in various instances, depending on whether only the ambient heat of the system is removed or whether some phase change with liberation or consumption of heat proceeds simultaneously. In Fig. 4.11, the cooling curves are shown schematically for various compositions in the phase diagram of a 2 — Ib_2 system. Cooling curves can be subdivided into several types:

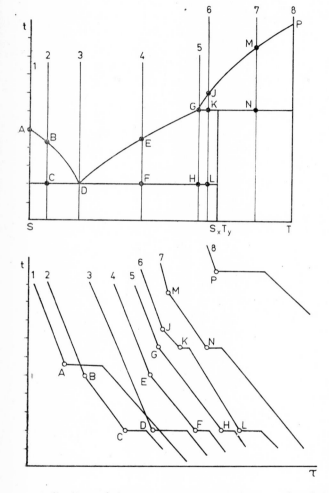

4.11. Cooling curves in a 2-Ib_2 type system.

(a) Cooling of the pure components (curves *1* and *8*). The first section represents cooling of the melt, point *A* or *P* lies at the beginning of a horizontal section representing solidification of the substance (the melting point) and then a section with negative slope follows for cooling of the solid phase.

(b) Cooling of a eutectic mixture (curve *3*). The first negatively sloped section represents cooling of the solution (melt), the horizontal section from point *B* corresponds to solidification of the eutectic mixture of substances S and T and the final negatively sloped section corresponds to cooling of the solid phases.

(c) **Cooling of solutions with non-eutectic composition** (curves *3, 4, 5, 6* and *7*). The first section again corresponds to cooling of the solution; the slope changes at points *B, E, G* and *M*, one of the solid phases separates and the heat of crystallisation is liberated; at points *C, F* and *H* both solid phases precipitate at the eutectic point; a peritectic change takes place on curve *7* at point *N;* all of these processes are isothermal. After the disappearance of the liquid phase, cooling of the mixture of solid phases takes place. On curve *6* the slope changes at point *J* and the heat of crystallisation is liberated during the separation of one solid phase; a peritectic change takes place at point *K* (at a constant temperature) and is then followed by separation of solid compound S_xT_y and finally, at point *L*, by simultaneous crystallisation of the eutectic at a constant temperature. Cooling of the mixture of solid phases then follows.

A different type of cooling curve is obtained with systems in which the components form solid solutions. A typical example is given in Fig. 4.12 for a 2-IIa$_1$

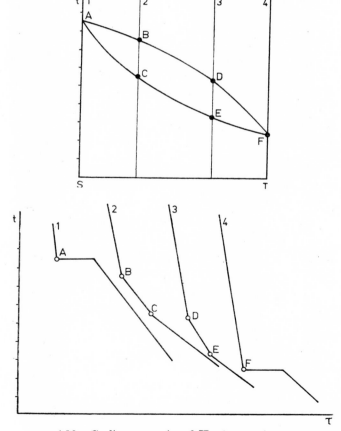

4.12. Cooling curves in a 2-IIa$_1$ type system.

type of system. The cooling curves can be classified into only two types in this instance.

(a) Cooling curves of the pure components (curves *1* and *4*), which are, of course, identical with those in the previous instance.

(b) Cooling curves of mixtures (curves *2* and *3*). After cooling of the solution, a sloped section at points B and D follows, corresponding to the precipitation of solid solutions with variable composition and at points C and E cooling of the solid phase (mixed crystals) follows. On these curves, no horizontal section appears (provided that this is not another type of diagram, such as 2-IIa_2 with composition corresponding to a maximum or minimum or 2-IIb with composition corresponding to a binary compound).

It is evident from the above examples that the whole phase diagram can be constructed on the basis of the breaks found on cooling curves.

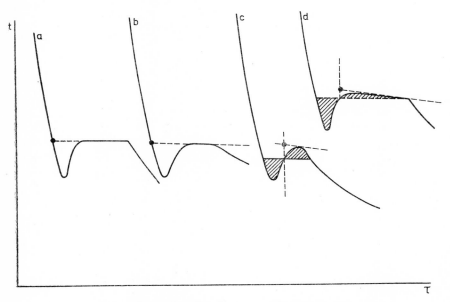

4.13. Correction of cooling curves for supercooling.

In reality, cooling curves are not as ideal as those shown in Figs. 4.11 and 4.12. During the above description of the synthetic methods, it was mentioned that the solution or melt can be supercooled during cooling. By separation of a solid phase from the supercooled liquid a greater amount of heat can be released and the temperature increases to the equilibrium value. If the cooling proceeds too rapidly or if the amount of heat released is too small, the measured cooling curve must be corrected. Examples of cooling curve corrections are shown in Fig. 4.13. It must be pointed out, however, that these corrections are empirical and if too large a correction is required (curve c), it is better either to repeat the experiment at a considerably lower cooling rate or to use another method for the determination of the phase equilibrium.

If the cooling curves are recorded in a broad temperature range, with the final temperature close to that of the surroundings, t_0, they are usually markedly curved owing to the decreasing temperature gradient and slower heat transfer. It is then better to plot the data in log $(t - t_0)$ versus time (τ) coordinates.

The cooling curve method is used chiefly for melts; it is rarely used with solutions, and then only for readily soluble substances with large heats of crystallisation.

4.3.2. The kinetics of the growth and dissolution of crystals

The difference in the behaviour of crystals immersed in undersaturated and supersaturated solutions was utilized in the Töppler method; in this method such state of the solution is required that the crystal neither grows nor dissolves. The rate of crystal growth can generally be described by an equation of the type

$$\frac{dg}{d\tau} = k_2 F(c - c_s)^n \qquad (4.5)$$

where dg is the increase in weight of the crystal with surface area F during a time interval $d\tau$, when the crystal is immersed in a supersaturated solution with a concentration of $c > c_s$. Frequently, especially in solutions that are not stirred too vigorously, the crystal growth is controlled by the diffusion rate and the exponent n is then unity. Similarly, the rate of crystal dissolution is given by the equation

$$\frac{dg}{d\tau} = -\frac{D}{\delta} F(c_s - c) \qquad (4.6)$$

where D is the diffusion coefficient, δ the thickness of the stationary layer at the crystal surface and $c < c_s$ the concentration of the undersaturated solution. On the basis of the above relationships, it is possible to find from the increase or decrease in weight of a crystal immersed in a series of solutions of various concentrations (or in a solution with a fixed concentration at various temperatures, and therefore with different values of c_s), the dependence of the crystal growth and dissolution rate on the concentration or temperature and to find by graphical extrapolation a solution state such that the crystal neither grows nor dissolves, i.e. the equilibrium state.

Basically two complications can be encountered when using this method:

(a) A layer of mother liquor adheres to the crystal after removal from the solution. If the procedure is reproducible, i.e., the crystal is always removed and dried in the same way (e.g., dried quickly with filter-paper and allowed to dry further in a desiccator), the correction for adhering mother liquor is roughly constant and can be determined by an independent experiment.

(b) Crystal dissolution is usually not the opposite process to growth and the behaviour of the system in the vicinity of the equilibrium point has not yet been explained theoretically. Several possible cases are depicted (considerably enlarged) in Fig. 4.14:

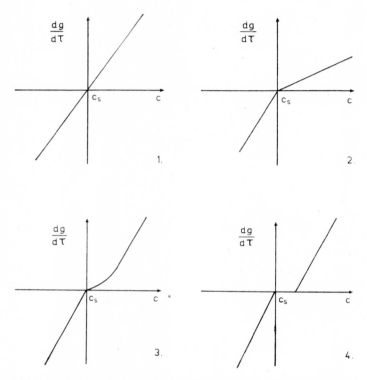

4.14. Crystal dissolution and growth in a region close to equilibrium.

1. growth and dissolution are governed solely by diffusion and are mutually reciprocal;
2. growth depends linearly on the supersaturation and generally proceeds at a different rate from dissolution;
3. growth starts non-linearly at small supersaturations and can exhibit a linear dependence only at higher supersaturations.
4. growth starts only at a certain supersaturation value. From Fig. 4.14 follow all the dangers that may be encountered during the extrapolation of experimental kinetic data (not taking into account poor precision of measurements). The extrapolation of the dependence of the rate of dissolution on the temperature of the solution is apparently relatively reliable. More care must be taken during the

extrapolation of crystal growth rates. It is recommended that the extrapolation should not be carried out in any instance by using one straight line, common for the regions of crystal growth and dissolution. An example of a graphical solution is shown in Fig. 4.15.

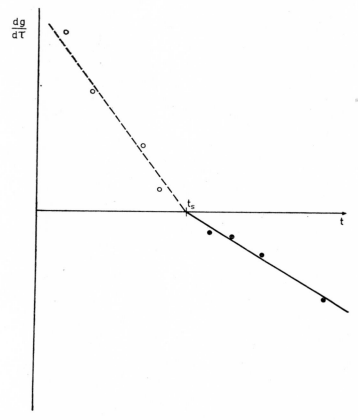

4.15. Example of the determination of the equilibrium temperature by the dynamic method

5. CORRELATION METHODS FOR CALCULATION OF PHASE EQUILIBRIA

It follows from the previous chapters that, in order to obtain data in multi-component systems, laborious and time consuming procedures are required; moreover, there are so many possible combinations of components that most of the phase equilibria in multicomponent systems have not yet been measured.

However, a method for handling the equilibrium data is useful even for those systems for which the pertinent results have already been published, as this method makes possible

(a) a check on the consistency of the published or measured values and

(b) interpretation of the data in the form of both equations and diagrams.

When measuring phase equilibria in multicomponent systems, the number of measurements necessary for the construction of a phase diagram rapidly increases with an increase in the number of components. Therefore, a data handling method that decreases the number of measurements to an acceptable level is desirable.

Methods that more or less comply with the above requirements for data correlation can be classified into two groups: purely empirical methods, based on certain (theoretically not supported) geometrical concepts, and methods derived from thermodynamic descriptions of phase equilibria, which replace unknown quantities by an empirical function. All of these methods permit transfer from systems with few components to more complex systems, performance of temperature interpolations (and also extrapolations to a limited extent), experimental data handling and construction of phase diagrams from a minimum number of values. Their application range and precision are, of course, determined by their character and by the number of initial values required; many of them can be used only for rough assessments unless corrections are introduced, while others yield data with a precision comparable to that attained during the experimental determination.

In the following sections, the most important correlation methods are surveyed.

5.1. Methods based on geometrical concepts

Methods belonging to this group employ the concepts illustrated in Fig. 3.15. The isotherms in a three-component system resemble one another to a certain extent and the line connecting the isotonic points passes through the eutectic point of the corresponding binary system. Hence, the isotherms in a three-component system could be simulated by using a suitably chosen projection of a binary phase diagram into the space occupied by the ternary system; in an analogous manner, it would be possible to transfer from three-component systems to four-component or even higher systems. The methods that are described differ in the projection procedure employed.

5.1.1. Orthogonal projection

Gromakov[1] derived relationships for the calculation of the properties of three-component and multicomponent systems from binary data, employing the arbitrary assumption that the properties of, e.g., ternary mixtures, are determined by the mutual relations of two of their components and are virtually independent of the third component.

In other words (see Fig. 5.1), if the solubility curve, σ_{AB}, of binary mixture A + B is projected in the direction of the y-axis so that the projection is determined by the shape of curve σ_{BC}, corresponding to binary mixture B + C, then solubility curve σ_{AC} should be unambiguously determined by the intercept with the plane constructed vertically above segment $\overline{AC}$.

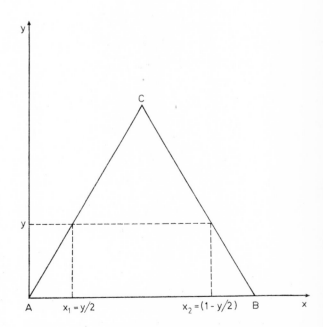

5.1. Principle of orthogonal projection.

[1] S. D. Gromakov, Zh. Fiz. Khim., *30* (1956) 2373; *30* (1956) 2621; *28* (1954) 1257; *31* (1957) 2597; *32* (1958) 232.

If the solubility in the ternary mixture is σ_{ABC} and y is constant, the change in the solubility caused by a change in composition is given by

$$\left(\frac{\partial \sigma_{ABC}}{\partial x}\right)_y = \varphi'(x). \tag{5.1}$$

On integrating this relationship, the equation

$$\sigma_{ABC}(x, y) = \int \varphi' \cdot (x) \, \mathrm{d}x + c_1(y) = \sigma_{AB}(x) + c_1(y) \tag{5.2}$$

is obtained. The value of the constant $c_1(y)$ is given by the condition for $x = 1$:

$$\sigma_{ABC}(1, y) = \sigma_{AB}(1) + c_1(y). \tag{5.3}$$

As it simultaneously holds that

$$\sigma_{AB}(1) = t_B \tag{5.4}$$

where t_B is the melting point of substance B, the basic equation representing the system is obtained:

$$\sigma_{ABC}(x,y) = \sigma_{AB}(x) + \sigma_{BC}(y) - t_B. \tag{5.5}$$

The above assumption cannot, of course, be generally valid. Therefore the projection is carried out in planes $y = $ constant along the two remaining curves, σ_{AC} and σ_{BC}, thus obtaining two planes; the resultant plane is given by a linear combination of these two planes, so that (again with reference to Fig. 5.1)

$$\sigma_1(x,y) = \sigma_{AB}(x) + \sigma_{BC}(y) - \sigma_{AB}(1-y/2) \tag{5.6}$$

$$\sigma_2(x,y) = \sigma_{AB}(x) + \sigma_{AC}(y) - \sigma_{AB}(y/2), \tag{5.7}$$

Linear combination of these two equations yields the relationship

$$\sigma_{ABC}(x, y) = \frac{x - y/2}{1 - y} \sigma_1(x, y) + \frac{1 - x - y/2}{1 - y} \sigma_2(x, y) =$$

$$= \sigma_{AB}(x) + \frac{x - y/2}{1 - y} [\sigma_{BC}(y) - \sigma_{AB}(1 - y/2)] +$$

$$+ \frac{1 - x - y/2}{1 - y} [\sigma_{AC}(y) - \sigma_{AB}(y/2)]. \tag{5.8}$$

For the constant value $y = y_0$, a series of functions is also constant:

$$\sigma_{AB}(x_1) = \sigma_{AB}(y_0/2) = a_1 \tag{5.9}$$

$$\sigma_{AC}(y) = \sigma_{AC}(y_0) = c_1 \tag{5.10}$$

$$\sigma_{AB}(x_2) = \sigma_{AB}(1 - y_0/2) = a_2 \tag{5.11}$$

$$\sigma_{BC}(y) = \sigma_{BC}(y_0) = c_2. \tag{5.12}$$

By substitution into the general eqn. 5.8 it follows that

$$\sigma_{ABC}(x,y) = \sigma_{AB}(x) + nx + m \tag{5.13}$$

where
$$n = \frac{(c_2 - a_2) - (c_1 - a_1)}{x_2 - x_1} \qquad (5.14)$$
and
$$m = (c_1 - a_1) - nx_1. \qquad (5.15)$$

The values of coefficients m and n can be found by solving the equations
$$nx_1 + m = c_1 - a_1 \qquad (5.16)$$
and
$$nx_2 + m = c_2 - a_2 \qquad (5.17)$$
where the values of a_1, a_2, c_1 and c_2 are defined by eqns. 5.9.— 5.12.

Eqn. 5.13 yields satisfactory results if eqn. 5.5 exhibits sufficiently small deviations from the experimental data. To check this statement, eqn. 5.5 can be rearranged to give
$$\sigma_{AC}(y) = \sigma_{AB}(y/2) + \sigma_{BC}(y) - t_B. \qquad (5.18)$$
If the σ_{AC} values calculated for a series of points by using functions σ_{AB} and σ_{BC} from eqn. 5.18 do not deviate from the real values by more than 5 %, calculation by means of eqn. 5.13 will be sufficiently accurate. Otherwise, the components must be replaced cyclically and that system chosen as the basic binary mixture, A + B, for which the deviations of eqn. 5.18 from reality are the smallest.

As an example, the calculation of several values of the melting points of nickel, copper and manganese alloys will be given; these points are denoted as *1*, *2* and *3* in Fig. 5.2. Fig. 5.2 also shows how the coordinates for the calculation are selected.

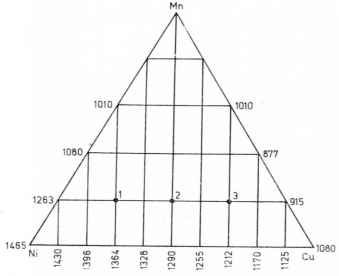

5.2. Example of calculation in the orthogonal projection method.

For points *1*, *2* and *3*, having the same y-coordinate, eqns. 5.16 and 5.17 are valid with the numerical values

$$0.1\,n + m = 1263 - 1430 = -167$$
$$0.9\,n + m = 915 - 1125 = -210$$
$$n = -53.8$$
$$m = -161.6.$$

The melting points corresponding to points *1*, *2* and *3* can be calculated from eqn. 5.13:

$$t_1 = 1364 - 0.3 \cdot 53.8 - 161.6 = 1186.2$$
$$t_2 = 1290 - 0.5 \cdot 53.8 - 161.6 = 1101.5$$
$$t_3 = 1212 - 0.7 \cdot 53.8 - 161.6 = 1012.7.$$

The calculated and experimental data agree to within 3—5 % in this case.

This method, which has been discussed in detail for a three-component system, can be extended to multicomponent systems. The calculation can be carried out successfully step by step or the whole calculation can be described by a single equation, which, however, is very complicated. The calculation procedure will be clarified by using a four-component system, depicted in Fig. 5.3. The melting point or the dissolution temperature at point G can be calculated as follows:

(a) By using known data for binary systems A + D, B + D and C + D, the values corresponding to points A_1, B_1 and C_1 are determined.

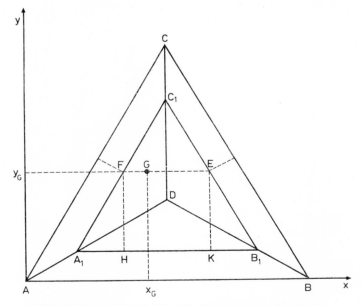

5.3. Orthogonal projection of a four-component system.

(b) The method described above for the calculation of ternary data is applied to three-component systems $A + B + D$, $A + C + D$ and $B + C + D$ and the values corresponding to points E, F, H and K are calculated.

(c) The above method for the calculation of ternary data is applied to the three-component system $A_1 + B_1 + C_1$, considering that

$$[A] + [B] + [C] + [D] = 1 \tag{5.19}$$

and hence the side of the triangle $A_1B_1C_1 = 1$ and the value at point G can be calculated.

The lever rule can be used to advantage in the calculation. For example, the value at point A_1 is calculated from the equation

$$[A_1] = \frac{x_D - x_{A1}}{x_D}[A] + \frac{x_{A1} - x_A}{x_D}[D] = (1 - z)[A] + z[D]. \tag{5.20}$$

Analogously, for example, for point E,

$$[E] = \frac{1 - y - z/2}{1 - z}[B_1] + \frac{y - z/2}{1 - z}[C_1] \tag{5.21}$$

and if the value at point G is to be expressed in terms of the values at points E and F, it then holds that

$$[G] = \frac{1 - x - y/2 - z/4}{1 - y - z/2}[F] + \frac{x - y/2 - z/4}{1 - y - z/2}[E]. \tag{5.22}$$

Application of equations of this type when calculating the properties of multi-component systems leads to such complicated equations that the use of the method (also considering the dubious initial assumption) becomes very questionable.

5.1.2. *Central projection*[1])

The description begins with the simplest case — the three-component system depicted in the upper part of Fig. 5.4. The values of the melting points or equilibrium temperatures are known in the whole concentration range for binary mixtures $A - B$, $A - C$ and $B - C$; the value of this quantity in the ternary system with the general composition corresponding to point S can then be determined. Assuming that there is no ternary interaction and hence the value at point S is additive with respect to suitably selected values in the binary mixtures, the following geometrical considerations can be employed.

It can be concluded on the basis of the general appearance of the phase diagram in Fig. 3.15 that eutonic points start in the binary eutectic points and that the line connecting them passes through a certain point in the area of the triangle.

[1] J. Nývlt, Chem. průmysl *9* (1959) 579; *10* (1960) 463.

If a solvent is one of the components, i.e., a substance whose melting point is relatively low, the ternary eutectic is located close to the point representing the solvent and the line connecting the eutonic points leads from the eutectic of the opposite binary mixture to the solvent apex. Therefore, the plane corresponding

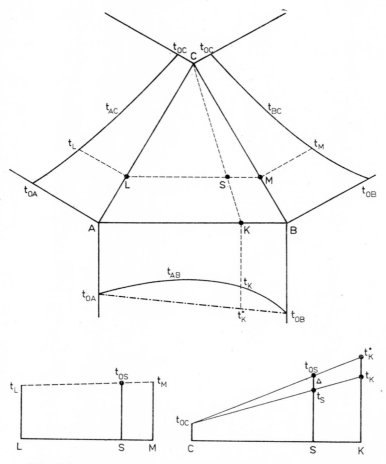

5.4. Scheme of projection from the point representing the solvent.

to the solubility of one of the components can be obtained by projecting the binary solubility curve (or the melting point curve) for the mixture with the other component from the apex corresponding to the solvent. The projection is carried out so that the projections of the terminal points (i.e., the melting points of the pure components) are identical with the two remaining binary curves. The projection procedure is best seen in Fig. 5.4. Through point S, at which the equilibrium saturation temperature or solubility of e.g., component B, is to be determined, two planes perpendicular to the plane of the triangle are constructed. One of the planes is parallel with the side corresponding to binary mixture A + B and the other passes through apex C.

The first plane, intersecting points L and M on the sides of the triangle with the corresponding t_L and t_M values, is shown at the bottom left-hand side of Fig. 5.4. The line connecting the t_L ant t_M values determines the value of t_{OS}, corresponding to additivity between the two boundary values and is then the projection of the additive value of t_{OA} and t_{OB}, i.e. of point t_K^*. The other plane is shown at the bottom right-hand side of Fig. 5.4; if a binary projection of points t_K^* and t_K is made in this plane into the coordinates of point S, the difference between the additive binary value, t_K^*, and the actual binary value, t_K, is projected as segment Δ. If segment Δ is then subtracted from the t_{OS} value found from the left-hand part of the figure, the actual t_S value is obtained, corresponding to the projection of the value at point K along curves t_{AC} and t_{BC}.

The above geometrical reasoning can then be expressed analytically. Firstly, the concentrations corresponding to the individual points in Fig. 5.4 will be determined. The molar of weight fraction of the ith component in the ternary mixture is denoted as x_i. The compositions corresponding to the individual points are then:

Point	[A]	[B]	[C]
S	x_A	x_B	x_C
K	$\dfrac{x_A}{x_A + x_B}$	$\dfrac{x_B}{x_A + x_B}$	0
L	$1 - x_C$	0	x_C
M	0	$1 - x_C$	x_C

The equilibrium temperatures are given by the relationships

$$t_K = t_{AB}\left[\frac{x_A}{x_A + x_B}\right]$$

$$t_L = t_{AC}[1 - x_C]$$

$$t_M = t_{BC}[1 - x_C]$$

$$t_{OS} = t_M + \frac{x_A}{x_A + x_B}(t_L - t_M) \qquad (5.23)$$

$$t_K^* = t_{OB} + \frac{x_A}{x_A + x_B}(t_{OA} - t_{OB}) \qquad (5.24)$$

$$\Delta = t_K + x_C(t_{OC} - t_K) - t_K^* - x_C(t_{OC} - t_K^*). \qquad (5.25)$$

Combination of eqns. 5.23 — 5.25 with the quation

$$t_S = t_{OS} - \Delta \tag{5.26}$$

results in the relationship

$$t_S = t_M + \frac{x_A}{x_A + x_B}(t_L - t_M) + x_A(t_K - t_{OA}) + x_B(t_K - t_{OB}). \tag{5.27}$$

Similar to orthogonal projection, the final equation is also non-symmetrical with respect to the components; hence the success of the method depends on the judicious selection of the basic binary mixture, A + B. When choosing the basic binary mixtures, some of the following empirical rules can be employed:

(a) The basic binary mixture is chosen so that it does not contain the substance with the lowest melting point (the solvent).

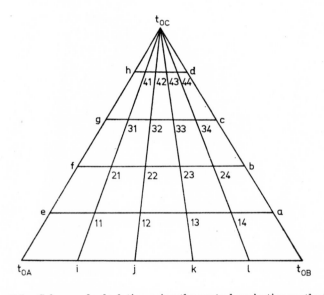

5.5. Scheme of calculation using the central projection method.

(b) If there are three comparable components in the system:

(i) binary mixtures are selected as A + C and B + C for which $t_{AC} + t_{BC}$ (at the same x_C value) is roughly a linear function of x_C;

(ii) a basic binary mixture is selected that has the minimum Δ value according to eqn. 5.25;

(iii) if at least one experimental point is known for the ternary mixture, the system with the smalles deviation at this point is used.

During the construction of the whole diagram, mechanized calculation can be used with advantage. The calculation is performed by using eqn. 5.27 for the points shown in Fig. 5.5 and the whole calculation is ordered in a table (Table 5.1). The calculation proceeds from one column to another and the points are numbered according to the columns and lines in Fig. 5.5. The numerical values in the table are rewritten during the calculation and complemented with the appropriate values for the binary mixtures; it then holds that

$$a = t_{BC} \quad [x_2 = 0.8]$$
$$b = t_{BC} \quad [x_2 = 0.6]$$
$$c = t_{BC} \quad [x_2 = 0.4]$$
$$d = t_{BC} \quad [x_2 = 0.2]$$
$$e = t_{AC} \quad [x_1 = 0.8]$$
$$f = t_{AC} \quad [x_1 = 0.6]$$
$$g = t_{AC} \quad [x_1 = 0.4]$$
$$h = t_{AC} \quad [x_1 = 0.2]$$
$$i = t_{AB} \quad [x_1 = 0.8]$$
$$j = t_{AB} \quad [x_1 = 0.6]$$
$$k = t_{AB} \quad [x_1 = 0.4]$$
$$l = t_{AB} \quad [x_1 = 0.2]$$

The products of the values between the double lines in Table 5.1 are then computed and finally the whole lines are summed. The calculated t_S values are plotted in the triangular diagram, rounded-off values are found by interpolations between neighbouring points and points with the same values are connected by continuous curves.

Table 5.1.

Point	t_M	$\dfrac{x_A}{x_A + x_B}$	$t_L - t_M$	x_A	$t_K - t_{OA}$	x_B	$t_K - t_{OB}$	t_S
11	a	0.80	$e - a$	0.64	$i - t_{OA}$	0.16	$i - t_{OB}$	
12	a	0.60	$e - a$	0.48	$j - t_{OA}$	0.32	$j - t_{OB}$	
13	a	0.40	$e - a$	0.32	$k - t_{OA}$	0.48	$k - t_{OB}$	
14	a	0.20	$e - a$	0.16	$l - t_{OA}$	0.64	$l - t_{OB}$	
21	b	0.80	$f - b$	0.48	$i - t_{OA}$	0.12	$i - t_{OB}$	
22	b	0.60	$f - b$	0.36	$j - t_{OA}$	0.24	$j - t_{OB}$	
23	b	0.40	$f - b$	0.24	$k - t_{OA}$	0.36	$k - t_{OB}$	
24	b	0.20	$f - b$	0.12	$l - t_{OA}$	0.48	$l - t_{OB}$	
31	c	0.80	$g - c$	0.32	$i - t_{OA}$	0.08	$i - t_{OB}$	
32	c	0.60	$g - c$	0.24	$j - t_{OA}$	0.16	$j - t_{OB}$	
33	c	0.40	$g - c$	0.16	$k - t_{OA}$	0.24	$k - t_{OB}$	
34	c	0.20	$g - c$	0.08	$l - t_{OA}$	0.32	$l - t_{OB}$	
41	d	0.80	$h - d$	0.16	$i - t_{OA}$	0.04	$i - t_{OB}$	
42	d	0.60	$h - d$	0.12	$j - t_{OA}$	0.08	$j - t_{OB}$	
43	d	0.40	$h - d$	0.08	$k - t_{OA}$	0.12	$k - t_{OB}$	
44	d	0.20	$h - d$	0.04	$l - t_{OA}$	0.16	$l - t_{OB}$	

It is important that proper functional dependences should be selected in the binary mixtures. For example, if A + B is a common eutectic system and if the isotherms of the solubility of substance A in a ternary mixture are calculated, then the branch corresponding to the solubility of substance A must be chosen as function t_{AB} and must be extrapolated over the whole range of the diagram. As only part of these data will play a role in the ternary mixture, errors that arise from extrapolation beyond the eutectic point will be unimportant.

The use of the method is demonstrated below on the determination of the melting points of alloys in the ternary mixture lead-bismuth-tin. The lead-bismuth binary mixture is chosen as the basis for the calculation and from the tabulated values for the individual binary mixtures the following quantities are calculated:

$$a = 200 \qquad i = 260$$
$$b = 140 \qquad j = 185$$
$$c = 173 \qquad k = 125$$
$$d = 210 \qquad l = 205$$
$$e = 278 \qquad t_{OA} = 322$$
$$f = 240 \qquad t_{OB} = 268$$
$$g = 181 \qquad t_{OC} = 232.$$
$$h = 200$$

Calculation of several points is shown in the following table:

Point	t_M	$\dfrac{x_A}{x_A + x_B}$	$t_L - t_M$	x_A	$t_K - t_{OA}$	x_B	$t_K - t_{OB}$	t_S
11	200	0.80	78	0.64	−62	0.16	−8	221
13	200	0.40	78	0.32	−197	0.48	−143	100
22	140	0.60	100	0.36	−137	0.24	−83	131
34	173	0.20	22	0.08	−117	0.32	−63	147
42	210	0.60	−10	0.12	−137	0.08	−83	176

All of the calculated values are plotted with the appropriate points and the straight lines are interpolated so as to obtain rounded-off values (Fig. 5.6). These rounded values are then connected by continuous curves. From the eutectic points of the binary mixtures, lines leading to the ternary eutectic are then obtained; these lines are connected as intercepts of corresponding isotherms and, if necessary, made more precise by calculating more points in the vicinity.

The method is applied below to a four-component system, represented by the tetrahedron given in Fig. 5.7. Three planes are constructed so as to pass through arbitrarily selected point S in the quaternary mixture. The first plane passes through points C, D and S; the second passes through points D and S and is parallel with side AB; the third plane, passing through point S, is parallel with the plane of the ternary mixture, A + B + C.

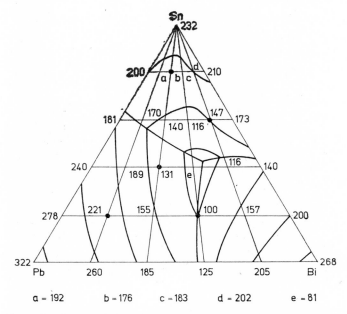

5.6. Example of construction of a diagram by the central projection method.

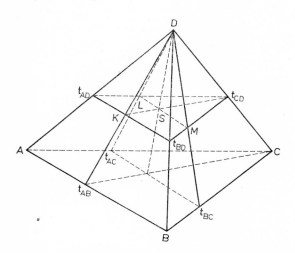

5.7. Application of central projection in a quaternary system.

The value at point S can be calculated by using eqn. 5.27 and triangle $t_{AD} - t_{BD} - t_{CD}$. The values at points K, L and M, required for the calculation, are obtained in an analogous manner from triangles ABD, BCD and ACD. Firstly, the composition of the system at important points is defined:

Point	[A]	[B]	[C]	[D]
t_{AB}	$\dfrac{x_A}{x_A + x_B}$	$\dfrac{x_B}{x_A + x_B}$	0	0
t_{BC}	0	$\dfrac{x_A + x_B}{1 - x_D}$	$\dfrac{x_C}{1 - x_D}$	0
t_{AC}	$\dfrac{x_A + x_B}{1 - x_D}$	0	$\dfrac{x_C}{1 - x_D}$	0
t_{AD}	$1 - x_D$	0	0	x_D
t_{BD}	0	$1 - x_D$	0	x_D
t_{CD}	0	0	$1 - x_D$	x_D
K	$\dfrac{x_A(1 - x_D)}{x_A + x_B}$	$\dfrac{x_B(1 - x_D)}{x_A + x_B}$	0	x_D
L	$\dfrac{x_A(1 - x_D)}{x_A + x_C}$	0	$\dfrac{x_C(1 - x_D)}{x_A + x_C}$	x_D
M	0	$\dfrac{x_B(1 - x_B)}{x_D + x_C}$	$\dfrac{x_C(1 - x_D)}{x_D + x_C}$	x_D
S	x_A	x_B	x_C	x_D

Applying eqn. 5.27 to calculation of the values inside the ternary diagrams, the following relationships are obtained:

$$t_K = t_{BD} + \frac{x_A}{x_A + x_B}(t_{AB} - t_{BD}) + \frac{1 - x_D}{x_A + x_B}[x_A(t_{AB} - t_{OA}) + x_B(t_{AB} - t_{OB})] \quad (5.28)$$

$$t_L = t_{AD} + \frac{x_C}{1 - x_D}(t_{CD} - t_{AD}) + (x_A + x_B)(t_{AC} - t_{OA}) + x_C(t_{AC} - t_{OC}) \quad (5.29)$$

$$t_M = t_{BD} + \frac{x_C}{1 - x_D}(t_{CD} - t_{BD}) + (x_A + x_B)(t_{BC} - t_{OB}) + x_C(t_{BC} - t_{OC}) \quad (5.30)$$

and

$$t_S = t_M + \frac{x_A}{x_A + x_B}(t_L - t_M) + \frac{x_A}{1 - x_D}(t_K - t_{AD}) + \frac{x_B}{1 - x_D}(t_K - t_{BD}) \quad (5.31)$$

or

$$t_S = \frac{x_A}{x_A + x_B} t_L + \frac{x_B}{x_A + x_B} t_M + x_A(t_{AB} - t_{OA}) + x_B(t_{AB} - t_{OB}). \quad (5.32)$$

By combining eqns. 5.28—5.32, the resultant equation for the calculation of the solubility in a four-component system on the basis of binary data is obtained:

$$t_S = (1 - x_D)\left[t_{BC} + \frac{x_A}{x_A + x_B}(t_{AC} - t_{BC})\right] + x_A(t_{AB} - t_{OA}) +$$

$$+ x_B(t_{AB} - t_{OB}) + \frac{1}{1-x_D}(x_A t_{AD} + x_B t_{BD} + x_C t_{CD}) -$$
$$+ (x_A t_{OA} + x_B t_{OB} + x_C t_{OC}). \tag{5.33}$$

If eqn. 5.27 for three-component systems is compared with eqn. 5.33 for four-component systems, a general relationship for n-component systems can be derived. For this purpose, a definition of deviations of binary values from additivity is introduced:

$$\Delta_{12} = t_{12} - \frac{x_1}{x_1 + x_2} t_{01} - \frac{x_2}{x_1 + x_2} t_{02} \tag{5.34}$$

$$\Delta_{13} = t_{13} - \frac{x_1 + x_2}{x_1 + x_2 + x_3} t_{01} - \frac{x_3}{x_1 + x_2 + x_3} t_{03} \tag{5.35}$$

etc.; in general,

$$\Delta_{in} = t_{in} - \frac{\sum_{i=1}^{n-1} x_i}{\sum_{i=1}^{n} x_i} t_{0i} - \frac{x_n}{\sum_{i=1}^{n} x_i} t_{on}, \tag{5.36}$$

Eqn. 5.27 for ternary systems can then be rewritten to give

$$t_S = x_1 t_{01} + x_2 t_{02} + x_3 t_{03} + \frac{x_1 + x_2}{x_1} \cdot x_1 \Delta_{12} +$$
$$+ \frac{x_1 + x_2 + x_3}{x_1 + x_2} (x_1 \Delta_{13} + x_2 \Delta_{23}). \tag{5.37}$$

For quaternary mixtures, eqn. 5.33 assumes the form

$$t_S = x_1 t_{01} + x_2 t_{02} + x_3 t_{03} + x_4 t_{04} + \frac{x_1 + x_2}{x_1} \Delta_{12} + \frac{x_1 + x_2 + x_3}{x_1 + x_2}(x_1 \Delta_{13} + x_2 \Delta_{23}) +$$
$$+ \frac{x_1 + x_2 + x_3 + x_4}{x_1 + x_2 + x_3}(x_1 \Delta_{14} + x_2 \Delta_{24} + x_3 \Delta_{34}). \tag{5.38}$$

Comparison of these two equations yields a general equation for n-component systems:

$$t_S = \sum_{i=1}^{n} x_i t_{0i} + \sum_{n=2}^{n} \left[\frac{\sum_{i=1}^{n} x_i}{\sum_{i=1}^{n-1} x_i} \left(\sum_{i=1}^{n} x_i \Delta_{in} \right) \right] \tag{5.39}$$

where Δ_{in} is defined by eqn. 5.36. This equation appears complicated but is actually only a comprised transcription of eqn. 5.38 including more terms.

Eqn. 5.39 can be used to calculate solubilities in a multicomponent system from the data on binary mixtures. As the equation for ternary systems, eqn. 5.27, was employed as the basis for its derivation, it is subject to roughly the same limitations. It can be expected that the precision of values assessed for multicomponent systems will be poorer than those for the simplest ternary system. However, as the method is based on the assumption of the additivity of binary data, deterioration of the precision will not be significant when no larger ternary interactions take place in the systems; interactions of higher orders are less probable. If stronger ternary interactions occur (which, of course, can usually by found only by determining the ternary data experimentally), it is more reliable to use the unsimplified eqn. 5.31, for calculations for four-component systems.

If few experimental data on a multicomponent system are available, an empirical correction can be made for the difference between the experimental and calculated data: $\Delta = \text{constant} \cdot x_1 x_2 x_3 \ldots x_n$.

Note: Our derivation of the orthogonal and central projection methods was performed for the calculation of phase equilibria. However, the methods can also be used for the calculation of other properties of multicomponent systems. The method can be applied successfully to thermodynamic and non-thermodynamic properties, equilibrium and non-equilibrium properties and to homogeneous and heterogeneous systems. The precision of the data often increases when only deviations from additivity are calculated.

The fact that the functional value (solubility) is calculated at a certain point in the phase diagram is a disadvantage of the method; it would often be more suitable to employ the reverse procedure, i.e., to determine the composition of a three-component mixture for the given solubility isotherm. This is made possible by a simple modification of the method:[1]

Eqn. 5.27 can be written in the form

$$(1 - x_\text{C}) t_K - \left[\frac{t_L - t_M}{1 - x_\text{C}} - t_{\text{OA}} + t_{\text{OB}} \right] x_\text{B} + t_L - t_S - t_{\text{OA}} + x_\text{C} t_{\text{OA}} = 0. \quad (5.40)$$

By selecting the isotherm, the t_S value is simultaneously determined. The t_L and t_M values are unambiguously determined by the selection of x_C, so that an equation with two unknown quantities remains. Eqn. 5.40 can thus be rewritten to give

$$\alpha t_K + \beta x_\text{B} + \gamma = 0 \quad (5.41)$$

where α, β and γ are constants at a constant x_C value and are given by the equations

$$\alpha = (1 - x_\text{C}) \quad (5.42)$$

$$\beta = - \left[\left(\frac{t_L - t_M}{1 - x_\text{C}} \right) - t_{\text{OA}} + t_{\text{OB}} \right] \quad (5.43)$$

and

$$\gamma = t_L - t_S - t_{\text{OA}} + x_\text{C} t_{\text{OA}}. \quad (5.44)$$

Eqn. 5.41 can be solved numerically or graphically.

During the numerical solution, the values of constants α, β and γ are calculated for the selected t_S and x_C values and an x_B value (and the corresponding t_K) is

[1] J. Nývlt and J. Gottfried, Chem. Průmysl, *14* (1964) 376.

sought by iterations such that eqn. 5.41 is fulfilled. The graphical solution is much easier. If the molar fraction of component B in a binary mixture A + B is written as

$$z_B = \frac{x_B}{1 - x_C} \qquad (5.45)$$

then eqn. 5.41 can be rearranged to give

$$t_K + \frac{\beta}{\alpha}(1 - x_C) z_B + \frac{\gamma}{\alpha} = 0 \qquad (5.46)$$

or (for comparison with eqn. 5.42)

$$t_K + \beta z_B + \frac{\gamma}{1 - x_C} = 0. \qquad (5.47)$$

As t_K is a function of z_B, the solution of eqn. 5.47 is given by the intercepts of the function

$$y = t_{AB}(z_B) \qquad (5.48)$$

with the straight line described by the equation

$$y = -\beta z_B - \frac{\gamma}{\alpha}. \qquad (5.49)$$

This straight line can easily be constructed from the conditions

$$z_B = 0: \quad y_A = -\frac{\gamma}{\alpha}. \qquad (5.50)$$

$$z_B = 1: \quad y_B = -\beta - \frac{\gamma}{\alpha}. \qquad (5.51)$$

Example: The 30 °C isotherm for the solubility of monochloroacetic acid (B) in a mixture of acetic acid (A) and dichloroacetic acid (C) is to be constructed. The initial data are given in Fig. 5.8 in which the phase diagram for A + B is shown, and in the first columns of the following table, summarizing the t_L and t_M values corresponding to systems A + C and B + C:

x_C	α	t_L	t_M	β	γ	$\frac{\gamma}{\alpha}$	$\left(-\frac{\gamma}{\alpha} - \beta\right)$	z_B	x_B	x_A
0.00	1.00	+16.0	+61.0	0.0	−30.0	30.0	30.0	—	—	—
0.10	0.90	+8.9	+56.7	+8.1	−35.9	39.9	31.8	0.715	0.644	0.256
0.20	0.80	−0.2	+51.0	+19.0	−43.3	54.2	35.2	0.766	0.613	0.187
0.30	0.70	−11.8	+44.3	+35.1	−53.3	76.2	41.1	0.835	0.585	0.115
0.40	0.60	−27.0	+35.9	+59.8	−66.8	111.4	51.6	0.929	0.557	0.043

The values for $x_C = 0$ are: $t_L = t_{OA}$, $t_M = t_{OB}$, $\beta = 0$, $\gamma = -t_S$. By using eqns. 5.42—5.44, the appropriate values of α, β and γ are calculated and from them, by using eqns. 5.50 and 5.51, the values of y_A and y_B are obtained. The lines connecting the points representing these values on axes drawn through points A and B intersect the phase equilibrium curve, t_{AB}, at points with coordinates z_B. By multiplying z_B by coefficient α, x_B is obtained and from it and from the originally selected x_C value, x_A is calculated.

The graphical solution is depicted in Fig. 5.8.

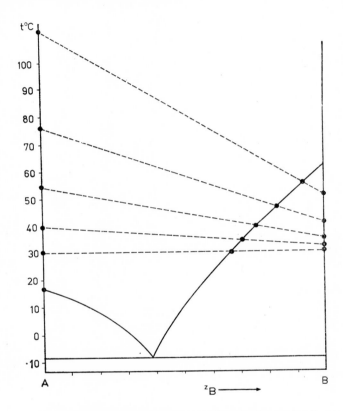

5.8. Graphical solution by the central projection method.

5.1.3. *A modified method of central projection*

In practice, it is often necessary to construct a phase diagram for a ternary system for which data on two binary mixtures are available and the shape of the function in the ternary mixture is known for a single value of a given parameter. As an example, the construction of the phase diagram can be considered in a system of two salts with a common ion and water at various temperatures, the dependences of the solubilities of the two pure salts in water and the phase diagram in the ternary mixture at a single temperature being known. Another example is monitoring the effect of the fourth component on the phase equilibrium in a system of two salts and water; here the effect of the fourth component on the solubilities of the individual salts and the ternary diagram of the system in the absence of the fourth component are usually known. For these systems the original method can be modified as follows[1]): The three-component system depicted in Fig. 5.9 is

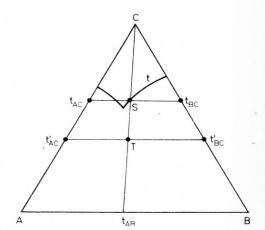

5.9. Scheme of a modified central projection method.

the starting point. The shape of one isotherm, t, and the temperature dependences of the solubilities of binary mixtures $A + C$ and $B + C$ are known in this system. The temperature at which a solution with composition given by point T will be just saturated can then be determined. Hence two planes perpendicular to the plane of the triangle and passing through point T can be constructed, one of them passing through apex C and the other parallel with side AB. The first plane intersects point $S(x_A, x_B, x_C)$ on isotherm t, for which eqn. 5.27 holds in the form

$$t_S = t_{BC} + \frac{x_A}{x_A + x_B}(t_{AC} - t_{BC}) + x_A(t_{AB} - t_{OA}) + x_B(t_{AB} - t_{OB}). \quad (5.52)$$

For point T, we can write in an analogous manner

$$t_T = t'_{BC} + \frac{x_A}{x_A + x_B}(t'_{AC} - t'_{BC}) + x'_A(t_{AB} - t_{OA}) + x'_B(t_{AB} - t_{OB}). \quad (5.53)$$

Solution of these two equations yields

$$t_T = t'_{BC} + \frac{x_A}{x_A + x_B}(t'_{AC} - t'_{BC}) + \frac{1 - x'_C}{1 - x_C}\left[t_S - \frac{x_A}{x_A + x_B}t_{AC} - \frac{x_B}{x_A + x_B}t_{BC}\right].$$

[1] J. Nývlt, J. Gottfried and L. Jäger, Chem. Průmysl 10 (1960) 341.

Eqn. 5.54 can be simplified and rewritten in the form (5.54)

$$t_T = G + \frac{1 - x'_C}{1 - x_C} F \tag{5.55}$$

where

$$F = t_S - \frac{x_A}{x_A + x_B} t_{AC} - \frac{x_B}{x_A + x_B} t_{BC} \tag{5.56}$$

is a function of $\dfrac{x_A}{x_A + x_B}$ alone, while the other term

$$G = t'_{BC} + \frac{x_A}{x_A + x_B} (t'_{AC} - t'_{BC}) \tag{5.57}$$

is a function of both $\dfrac{x_A}{x_A + x_B}$ and x'_C.

The whole computation can be mechanized if it is carried out for points located on the straight lines $x'_C =$ constant and $\dfrac{x_A}{x_A + x_B} =$ constant. The most lucid procedure again employs a table, which can be arranged in the form given in Table 5.2. The computation procedure is completely analogous to that followed in the original central projection method. The table is completed, one column at a time, from left to right; the G and F values for the individual lines equal the sum of the products of the values between the double lines. Finally, the t_T values are obtained, plotted as coordinates for the appropriate points in the diagram, the rounded-off values are found by interpolation from the neighbouring point and the identical values are interconnected by isotherms.

Table 5.2. Scheme of the computation using the modified central projection method

Point No.	t'_{BC}	$\dfrac{x_A}{x_A + x_B}$	$t'_{AC} - t_{BC}$	G	t_S	$\dfrac{-x_A}{x_A + x_B}$	t_{AC}	$\dfrac{-x_B}{x_A + x_B}$	t_{BC}	F	$\dfrac{1 - x'_C}{1 - x_C}$	t_T

Note: The value of function t_S was again taken to be the temperature at which the solution becomes saturated. Analogously, this function may be any other parameter mentioned in section 5.1.2; moreover, the concentration of the fourth component can also be chosen as parameter t, of course at constant temperature. This leads to a very common application of the method, namely, the case when the solubilities of substances A and B in ternary system A + B + C are known, as well as the effect of substance D on the solubility of A in system A + C and on the solubility of substance B in system B + C at the same temperature. As a result of the computation an isothermal phase diagram of the fourcomponent system is obtained in the form of a triangular diagram with apices A, B and C + D, the content of substance D being represented by coordinates written on the individual calculated solubility curves.

In a similar manner to the simple method of central projection, this modification also requires the numerical calculation of a whole series of points and then interpolation among these points. Therefore, this method was modified into a graphical determination of the points corresponding to a certain isotherm.[1] The modified central projection method can also be solved graphically. For this purpose, eqn. 5.54 is rewritten in the form

$$y_1(x'_C) = y_2(x'_C) \tag{5.58}$$

where

$$y_1(x'_C) = \frac{x_A}{x_A + x_B} t'_{AC} + \frac{x_B}{x_A + x_B} t'_{BC} \tag{5.59}$$

$$y_2(x'_C) = t_T - \frac{1 - x'_C}{1 - x_C} \left[t_S - y_1(x_C) \right] \tag{5.60}$$

$$y_1(x_C) = \frac{x_A}{x_A + x_B} t_{AC} + \frac{x_B}{x_A + x_B} t_{BC}. \tag{5.61}$$

If a straight line with a given constant ratio $x_A/(x_A + x_B)$ is chosen in the ternary diagram (Fig. 5.9), it intersects isotherm t at a point with coordinate x_C. We can then find a point T with unknown coordinate x'_C on this straight line with constant $x_A(x_A + x_B)$ which lies on isotherm t'. The coordinates of this point must correspond to the solution of eqns. 5.58—5.61.

The equation of $y_1(x'_C)$ or $y_1(x_C)$ is generally non-linear. If, for a number of arbitrarily chosen values of x_C and x'_C, the corresponding t_{AC} and t_{BC} or t'_{AC} and t'_{BC} values, respectively, are known, the $y_1(x_C)$ dependence corresponding to constant $x_A/(x_A + x_B)$ can be readily constructed graphically. This construction is apparent from Fig. 5.10. On the right-hand side of Fig. 5.10. are depicted the appropriate dependences using known data for $t_{AC}(x_C)$ and $t_{BC}(x_C)$. At point x_C, corresponding to point S in Fig. 5.9, a vertical straight line is constructed, intersecting solubility curves t_{AC} and t_{BC} at points P_2 and R_2, respectively. These points are transferred to the corresponding axes in the left-hand half of Fig. 5.10, giving points P_1 and R_1, respectively. The line connecting the latter two points gives dependence $y_1(x_C)$ as a function of $x_A/(x_A + x_B)$. At coordinate $x_A/(x_A + x_B)$, corresponding to point S in Fig. 5.9, point Q_1 is found and transferred back to

---

[1] J. Nývlt and Z. Blechta, Chem. Průmysl, 18 (1968) 371.

the right-hand half of Fig. 5.10, thus giving point Q_2. In an analogous manner, other points of dependence $y_1(x_C)$ are obtained, e.g., point Q_2' from points P_2' and R_2', and a curve is constructed through these points.

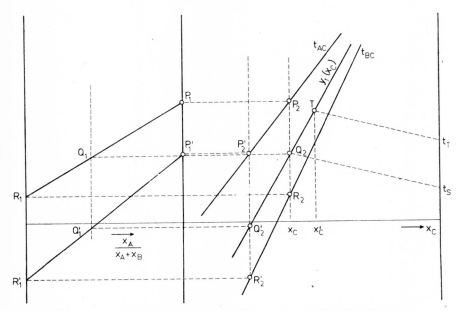

5.10. Graphical solution by the modified central projection method.

Eqn. 5.60 corresponds to a straight line; in its construction, two boundary conditions can be used:

$$\text{(a)} \quad x_C' = 1 \tag{5.62}$$

$$y_2(1) = t_T$$

$$\text{(b)} \quad x_C' = x_C$$

$$y_2(x_3) = y_1(x_3) + t_T - t_S. \tag{5.63}$$

In Fig. 5.10, a connecting line is constructed from points t_S with coordinate $x_C = 1$ to point Q_2 and a parallel line is drawn from point t_T with coordinate $x_C = 1$. This parallel line intersects curve $y_1(x_C)$ at point T with the coordinate to be determined, x_C'.

Example: The effect of the presence of sodium chloride (D) on the phase equilibrium in the system sodium carbonate (A)-sodium hydrogen carbonate (B)-water (C) is to be found. Data for the construction of the ternary diagram for sodium carbonate-sodium hydrogen carbonate-water at 30 °C have been published in the literature:

Point	$\dfrac{x_A}{x_A + x_B}$	x_C
0	1.000	71.5
1 (E)	0.967	71.0
2	0.900	76.0
3 (E)	0.7825	78.0
4	0.600	85.0
5	0.200	89.0
6	0.000	90.0

The literature also provides data on the phase equilibria of sodium carbonate–sodium chloride–water and sodium hydrogen carbonate–sodium chloride–water at the same temperature.

The following values have been calculated from the data:

t_{AC}	x_C	t_{BC}	x_C
0	71.5	0	90.0
5	75.0	5	93.0
10	78.5	10	96.0
15 (E)	82.0	15	97.5
20	90.0	20	98.2
25	98.0	25	98.7
27.2	100.0	27.2	100.0

The t_{AC} and t_{BC} values denote the sodium chloride content in weight per cent. The t_{AC} and t_{BC} values are plotted as functions of x_C in the right-hand half of Fig. 5.11, where all the constructions described in the previous sections were carried out for all of points 0—6. The x_C coordinates found were plotted on the corresponding straight lines, $x_A/(x_A + x_B)$, in the diagram given in Fig. 5.12, where several experimental points, determined by the analytical method, are also plotted for the sake of comparison.

Another modification of the method of central projection is also derived from eqn. 5.27, based on the selection of values of function t as the reduced concentrations related to the concentration of component B in the saturated solution, which is equal to unity:

$$t_{AB} = \frac{x_B}{x_A + x_B} \cdot \frac{1}{(x_B)_{\text{sat}(AB)}} \qquad [A + B] \qquad (5.64)$$

$$t_{AC} = \frac{x_C}{(x_C)_{\text{sat}(AC)}} \qquad [A + C] \qquad (5.65)$$

128

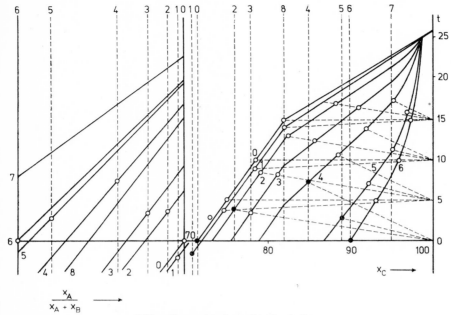

5.11. Example of graphical solution.

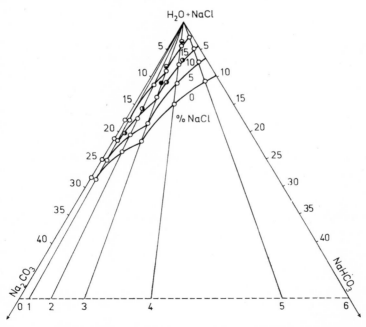

5.12. Phase equilibria in the system Na$_2$CO$_3$ (A)-NaHCO$_3$ (B)-NaCl (D)-H$_2$O (C) at 30 °C
Calculated points: ○, 11 % NaCl
Experimental points: ●, 11 % NaCl, ◐, 10 % NaCl, ⊖, 15 % NaCl.

$$t_{OA} = \frac{x_B}{(x_B)_{\text{sat}(AB)}} = 0 \qquad [x_B = 0] \qquad (5.66)$$

$$t_{OB} = \frac{1}{(x_B)_{\text{sat}(AB)}} \qquad [x_B = 1]. \qquad (5.67)$$

If t_{BC} is chosen as the unknown value, eqn. 5.27 can be rewritten in the form

$$t_S = \frac{x_B}{x_A + x_B} t_{BC} + \frac{x_A}{x_A + x_B} \cdot \frac{x_C}{(x_C)_{\text{sat}(AC)}}. \qquad (5.68)$$

of a point on an isotherm is to be determined, it must hold that

$$t_S = 1 \qquad (5.69)$$

hence

$$t_{BC} = \frac{x_A}{x_B} \left(1 - \frac{x_C}{(x_C)_{\text{sat}(AC)}}\right) + 1. \qquad (5.70)$$

The right-hand side of eqn. 5.70 corresponds to a straight line, for the boundary points of which it must hold that

$$t_{BC} = \frac{x_A}{x_B} + 1 \qquad [x_C = 0] \qquad (5.71)$$

$$t_{BC} = 1 \qquad [x_C = (x_C)_{\text{sat}(AC)}]. \qquad (5.72)$$

Hence if families of straight lines eqn. (5.70) for various x_A/x_B values are plotted on the graph in which individual isotherms, t_{BC}, are plotted as functions of x_C, the intercepts of these straight lines with the appropriate isotherms, t_{BC}, determine the x_C value and hence directly a point on the isotherm in the ternary diagram. The construction of various isotherms $t_{BC}(x_C)$ remains to be solved. From the known shape of one isotherm in the ternary diagram, the appropriate t_{BC} curve as a function x_C can be constructed for the same temperature, using eqn. 5.70. It is then possible to construct an auxiliary plot of the dependence of function t_{BC} on temperature for various values of constant x_C: for $x_C = 0$, the curve can be constructed in the whole required temperature range and linearized by a suitable selection of the temperature scale. On plotting functions $\log (t_{BC})_{x_C=\text{const}}$ versus $1/T$, a family of straight lines should be obtained that can be extrapolated to even higher temperatures for $x_C = 0$. The values for the construction of graph $t_{BC}(x_C)$ are then read from this auxiliary plot.

Example: The 50 °C isotherm is to be constructed in ternary system C — B — water if the temperature dependence of the solubility in systems C — water and D — water and the 20 °C isotherm in ternary mixture C — B — water are given (see Fig. 5.13). The calculation of auxiliary values is summarized in the table below. The first column gives the temperature, the second the selected x_C value, the third the x_C value divided by the $(x_C)_{\text{sat}(C-H_2O)}$ value at 50 °C, i.e., 0.5, the fourth the x_B

values corresponding to the given isotherm and the selected x_C values, the fifth the $x_{H_2O} = x_A$ values, calcutated by subtraction of the former values from unity, the sixth fractions x_A/x_B and the seventh the t_{BC} value calculated from the data given in the third and fourth columns by using eqn. 5.70.

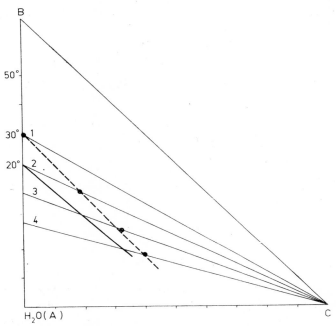

5.13. Alternative modification of the central projection method.

1	2	3	4	5	6	7	8
t °C	x_C	$\dfrac{x_C}{(x_C)_{sat}}$	x_B	x_A	x_A/x_B	$(t_{BC})_{20\,°C}$	$(t_{BC})_{50\,°C}$
20	0.000	0.000	0.500	0.500	1.000	2.000	1.666
	0.050	0.100	0.450	0.500	1.111	2.000	1.666
	0.100	0.200	0.405	0.495	1.222	1.978	1.655
	0.150	0.300	0.355	0.495	1.394	1.976	1.650
	0.200	0.400	0.315	0.485	1.540	1.924	1.618
	0.250	0.500	0.270	0.480	1.770	1.889	1.592
	0.300	0.600	0.225	0.475	2.111	1.844	1.564
	0.350	0.700	0.185	0.465	2.514	1.754	1.500
50	0.000	0.000	0.600	0.400	0.666	1.666	

The t_{BC} values are transferred onto the graph (the bottom part of Fig. 5.14) representing the dependence of $t_{BC}(x_C)$ at 20 °C. In the upper part of Fig. 5.14 the temperature extrapolation using the t_{BC} values at 20 and 50 °C and for $x = 0$ is depicted schematically. If more precise data are not available, the following approximation is employed:

$$\frac{(t_{BC})_{50} - 1}{(t_{BC})_{20} - 1} = \text{constant}. \tag{5.73}$$

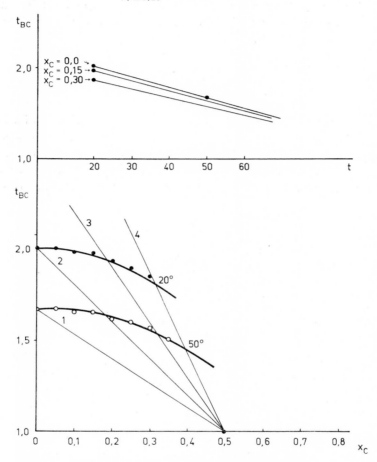

5.14. Auxiliary graphs for the construction depicted in Fig. 5.13.

The extrapolated values were calculated and are given in the eighth column of the above table and represented graphically in the lower part of Fig. 5.14. A family of straight lines is then constructed through point B in Fig. 5.13 for various x_A/x_B ratios: the lines shown in the figure correspond to $x_A/x_B = 0.666$, 1.000, 1.500 and 2.333. By using eqns. 5.71 and 5.72, the values of the boundary points of the straight lines drawn in the bottom part of Fig. 5.14 were calculated from these values. The intercepts of these straight lines with isotherm t_{BC} at 50 °C yield the x_C values corresponding to points on the 50 °C isotherm. If the x_C values are plotted on the straight lines of the corresponding x_A/x_B ratios in Fig. 5.13, points on the 50 °C isotherm are obtained. In an analogous manner (by exchanging symbols B and C), the other branch of the isotherm can be constructed.

The method described is simple and very rapid. However, it has the disadvantage of the necessity for the temperature extrapolation of the t_{BC} dependence; hence, it is more suitable for problems that involve interpolation between two known isotherms in a ternary system.

5.1.4. *Clinogonial projection*

The construction of ternary diagrams in the methods described so far involved projection of the t_{AB} functional dependence on to a plane with x_C = constant. If the known shape of one isotherm in a ternary mixture is to be re-calculated to another value of the temperature (or another parameter), it is necessary to know the solubility dependence studied in binary systems $A + C$ and $B + C$ over

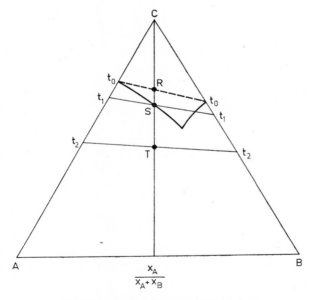

5.15. Principle of clinogonial projection.

a very wide range, especially when the ternary diagram is markedly non-symmetrical, i.e., when the behaviours of the two binary systems are very different. This disadvantage is overcome by the following rapid method.[1]

[1] J. Nývlt and Z. Blechta, Chem. Průmysl, *18*, (1968) 426.

A three-component system A + B + C, represented in Fig. 5.15, is considered. The quantity monitored (e.g. the temperature corresponding to a saturated solution) generally has t_{AB}, t_{AC} and t_{BC} values in the corresponding binary systems. A number of planes are constructed at right angles to triangle ABC, so that they pass through the same coordinate in binary systems AC and BC, i.e., $t_{AC} = t_{BC}$. Another system of planes perpendicular to triangle ABC passes through point C. The required value of the saturation temperature at point S, corresponding to $t_S - t_0$, is given, as in the previous methods, by the t_1 value, corrected for the projection of the difference of the appropriate t_{AB} value from the linear dependence, i.e.

$$t_S = t_1 + \Delta t_{AB}(1 - x_C(S)) \tag{5.74}$$

where

$$\Delta t_{AB} = t_{AB} - t^*_{AB} \tag{5.75}$$

$$t^*_{AB} = t_{OB} + \frac{x_A}{x_A + x_B}(t_{OA} - t_{OB}). \tag{5.76}$$

On substitution into eqn. 5.74, the resulting relationship

$$t_S = t_1 + (1 - x_C(S))\left(t_{AB} - \frac{x_A}{x_A + x_B}(t_{OA} - t_{OB}) - t_{OB}\right) \tag{5.77}$$

is obtained. The calculation according to eqn. 5.77 is very simple and the whole procedure can again be mechanized by assembling a suitable table.

Very frequently, one isotherm in a ternary diagram is known rather than the shape of function t_{AB}; thus the shape of curve t_0 in Fig. 5.15 is known and the shape of another isotherm can be determined. On the basis of eqn. 5.74 and analogous eqn. 5.78, written for an arbitrary point, T, located on the same straight line, $x_A/(x_A + x_B)$,

$$t_T = t_2 + \Delta t_{AB}(1 - x_C(T)) \tag{5.78}$$

the relation

$$\frac{t_S - t_1}{1 - x_C(S)} = \frac{t_T - t_2}{1 - x_C(T)} = \Delta t_{AB} \quad \left[\frac{x_A}{x_A + x_B} = \text{const.}\right] \tag{5.79}$$

can then be written. Eqn. 5.79 expresses the linear dependence between t_2 and $x_C(T)$. The other dependence between t and x_C is obtained directly from the ternary diagram. For example, the following corresponding pairs of values can be read directly from Fig. 5.15: $x_C(R) \ldots t_0$; $x_C(S) \ldots t_1$; $x_C(T) \ldots t_2$ etc. Both dependences are depicted schematically in Fig. 5.16. First, the $t(x_C)$ dependence is plotted on the graph, i.e., the x_C values corresponding to the intercepts of various connecting lines, $t_{AC} = t_{BC}$, with straight line $x_A/(x_A + x_B)$ are found from the triangular diagram. Further, the t_S value is plotted on the graph on the vertical line constructed through point $x_C = 1$. The line connecting point t_S with point t_1, located on the vertical line passing through point $x_C = x_C(S)$, determines the length of segment $\overline{KL} = \Delta t_{AB}$. An arbitrary t_T value is selected and the appropriate segment is plotted on the vertical line passing through point $x_C = 1$, a line is drawn parallel to straight line $t_S L$ and the intercept of this line with curve $t(x_C)$ yields the $x_C(T)$ value, corresponding to the t_T isotherm in Fig. 5.15.

Example: The isotherm of the ammonium sulphite solubility curve at 20 °C in the system ammonium sulphite-ammonium sulphate-water is drawn in Fig. 5.17,

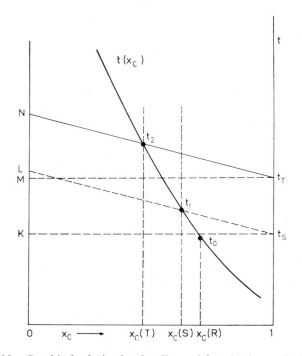

5.16. Graphical solution by the clinogonial projection method.

and the points representing aqueous solutions of ammonium sulphite and ammonium sulphate saturated at 0, 20, 50 and 70 °C are also given. For an arbitrarily chosen straight line a, the x_{H_2O} values corresponding to the intercepts of this line with dashed connecting lines for identical temperatures in the binary mixtures are read from this diagram. The $t(x_{H_2O})$ dependence found in this manner is plotted in auxiliary graph 5.18 (the points denoted by open circles). From coordinate $t = 20$ °C at $x_{H_2O} = 1$, a connecting line is constructed to the point on curve $t(x_{H_2O})$ located at coordinate $x_{H_2O} = 0.59$, corresponding to the intercept of the 20 °C isotherm with straight line a in Fig. 5.17. Lines parallel to this connecting line are then drawn in Fig. 5.18 from coordinates $t = 0$, 50 and 70 °C and their intercepts (closed circles) with curve $t(x_{H_2O})$ are found. The x_{H_2O} values obtained are then transferred back on to the triangular diagram given in Fig. 5.17 (points denoted by closed circles). The procedure for other straight lines corresponding to other salt ratios is analogous.

The method described is very rapid and yields satisfactory results unless the extrapolations are carried out over too wide a range.

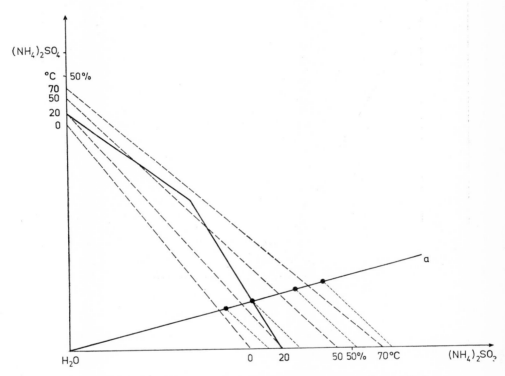

5.17. Example of construction by the clinogonial projection method.

5.1.5. *The symmetrical correlation method*

The correlation methods so far described are non-symmetrical with respect to the components, i.e., one of the partial binary systems is considered as the basic system. This condition is readily applicable to those three-component systems where one of the components can be considered to be a solvent with properties sufficiently different from those of the remaining two components. If, however, all three components have similar properties, it is difficult to choose the basic binary mixture. It is then more convenient to employ a method symmetrical with respect to all three components, i.e., a method that yields a correlation result independent of the order in which the components are chosen.

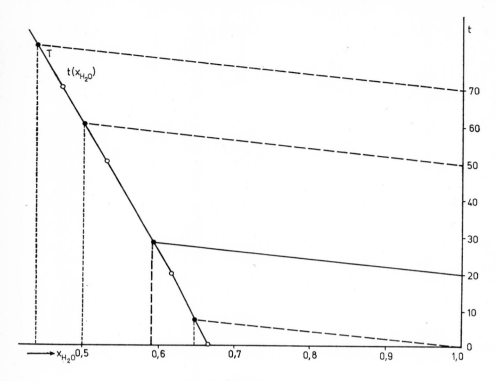

5.18. An auxiliary graph for Fig. 5.17.

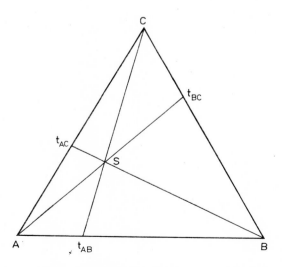

5.19. Scheme of the symmetrical correlation method.

One of the simplest modifications of a symmetrical correlation method[1] is based on the scheme depicted in Fig. 5.19. The coordinates of the individual points in the diagram are:

Point	[A]	[B]	[C]
S	x_A	x_B	x_C
A	1	0	0
B	0	1	0
C	0	0	1
t_{AB}	$\dfrac{x_A}{x_A + x_B}$	$\dfrac{x_B}{x_A + x_B}$	0
t_{AC}	$\dfrac{x_A}{x_A + x_C}$	0	$\dfrac{x_C}{x_A + x_C}$
t_{BC}	0	$\dfrac{x_B}{x_B + x_C}$	$\dfrac{x_C}{x_B + x_C}$

In an ideal case, characterized by additivity of the properties in the ternary mixture, the relationship

$$t_S^* = x_A\, t_{OA} + x_B\, t_{OB} + x_C\, t_{OC} \tag{5.80}$$

can be written. The actual t_S value will differ from the ideal value by a certain amount, Δ_{ABC}

$$t_S = t_S^* + \Delta_{ABC} \tag{5.81}$$

the magnitude of which is determined by the projection of the corresponding binary deviations onto point S. If, for the deviation value in a binary mixture,

$$\Delta_{ij} = t_{ij} - \frac{x_i}{x_i + x_j} t_{oi} - \frac{x_i}{x_i + x_j} t_{oj} \tag{5.82}$$

where subscripts i and j denote the corresponding components of binary mixture $A + B$, $A + C$ and $B + C$, and the equation

$$\Delta_{ABC} = (1 - x_C)\, \Delta_{AB} + (1 - x_B)\, \Delta_{AC} + (1 - x_A)\, \Delta_{BC} \tag{5.83}$$

for the ternary deviation can be written, then the resulting relationship is obtained in the form

$$t_S = (1 - x_C)\, t_{AB} + (1 - x_B)\, t_{AC} + (1 - x_A)\, t_{BC} - x_A\, t_{OA} - x_B\, t_{OB} - x_C\, t_{OC}. \tag{5.84}$$

As already pointed out, eqn. 5.84 is suitable for the correlation of ternary data in systems in which the melting points of the three components are similar.

---

[1] J. Nývlt, Chem. Průmysl, *18* (1968) 260.

So far, the real melting points of pure components A, B and C have been used in the calculations. If, for example, the solubility of substance A in ternary mixture A + B + C is followed, it was assumed that only that portion of binary dependence t_{AB} which represented the region of phase equilibrium of the solution or melt with pure salt A played a role in the calculation, i.e. that the portion adjacent to apex B was unimportant. However, this assumption need not be generally valid and the solubility curve for substance A in binary system AB would have to be extrapolated beyond the eutectic point or even over the whole range of the binary mixture. Such an extrapolation is represented schematically in Fig. 5.20. If it is assumed that the shape of the extrapolated dependence, $t_{A(B)}t_{A(C)}$,

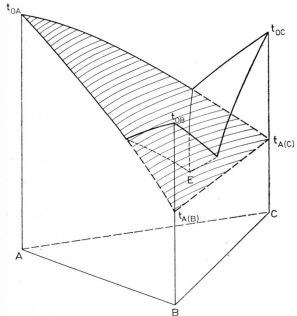

5.20. Scheme of the symmetrical corelation method with cyclic components replacement.

has no great importance in the calculation of the behaviour of the system close to point A, then it can be considered to be linear[1]. If eqns. 5.82 and 5.83 are then applied to the calculation of the behaviour of the system, $\Delta_{BC} = 0$ and combination of the two equations yields the relationship

$$\Delta_{ABC} = (1 - x_C)\left(t_{AB} - \frac{x_A}{x_A + x_B}t_{0A} - \frac{x_B}{x_A + x_B}t_{A(B)}\right) +$$
$$+ (1 - x_B)\left(t_{AC} - \frac{x_A}{x_A + x_C}t_{0A} - \frac{x_C}{x_A + x_C}t_{A(C)}\right). \quad (5.85)$$

[1] J. Čársky, Chem. Zvesti, 25 (1971) 266.

As it is simultaneously valid that

$$x_A + x_B + x_C = 1 \tag{5.86}$$

it follows that

$$\Delta_{ABC} = (1 - x_C) t_{AB} + (1 - x_B) t_{AC} - x_A t_{AC} - x_A t_{OA} - x_B t_{A(B)} - x_C t_{A(C)}. \tag{5.87}$$

On substitution into eqn. 5.81 and using eqn. 5.80, the following relationship is obtained:

$$t_{S(A)} = (1 - x_C) t_{AB} + (1 - x_B) t_{AC} - x_A t_{OA}. \tag{5.88}$$

Cyclic replacement then gives

$$t_{S(B)} = (1 - x_A) t_{BC} + (1 - x_C) t_{AB} - x_B - x_B t_{OB} \tag{5.89}$$

$$t_{S(C)} = (1 - x_B) t_{AC} + (1 - x_A) t_{BC} - x_C t_{OC}. \tag{5.90}$$

It is important to note that extrapolated, and thus imprecise, values of $t_{A(B)}$, $t_{A(C)}$ and the corresponding values for the other two components are not present in eqns. 5.88—5.90.

The shape of the isotherms over the whole diagram are calculated as follows. The calculations for regions close to apices A, B and C are carried out employing eqns. 5.88, 5.89 and 5.90, respectively. Basically, three fields of points are thus obtained and interpolation is carried out among them by the procedure described above and the isotherms are then constructed. The intercepts of the corresponding isotherms are connected with the appropriate binary eutectic; three such connecting lines intersect one another at the ternary eutectic.

Example: The temperature is to be calculated at which solution S with composition $x_A = 0.73$, $x_B = 0.09$ and $x_C = 0.18$ is saturated with component A, the behaviour of binary systems $A + B$ and $A + C$ being depicted in Fig. 5.21. The solubility of substance A can be calculated from eqn. 5.88. The t_{AB} value for $x_A/(x_A + x_B) = 0.89$ and the t_{AC} value for $x_A/(x_A + x_C) = 0.80$ are found from Fig. 5.21: $t_{AB} = 85.5$, $t_{AC} = 70.5$. These values, together with $t_{OA} = 89.5$, are substituted into eqn. 5.88:

$$t_{S(A)} = 0.82 \cdot 85.5 + 0.91 \cdot 70.5 - 0.73 \cdot 89.5 = 68.93.$$

Similarly, for point $T(x_A = 0.77; x_B = 0.10; x_C = 0.13)$, the relations

$$t_{AB}\left[\frac{x_A}{x_A + x_B} = 0.89\right] = 85.5,$$

$$t_{AC}\left[\frac{x_A}{x_A + x_C} = 0.86\right] = 76.0$$

and

$$t_{T(A)} = 0.87 \cdot 85.5 + 0.90 \cdot 76.0 - 0.77 \cdot 89.5 = 73.87$$

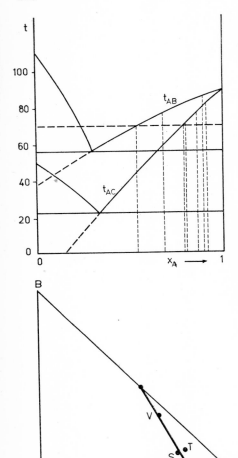

are obtained. Finally, for the third point, $V(x_A = 0.635; x_B = 0.30; x_C = 0.065)$ it is found that

$$t_{AB}\left[\frac{x_A}{x_A + x_B} = 0.68\right] = 77.0;$$

$$t_{AC}\left[\frac{x_A}{x_A + x_C} = 0.91\right] = 81.0$$

and

$$t_{V(A)} = 0.935 \cdot 77.0 + 0.70 \cdot 81.0 - \\ - 0.635 \cdot 89.5 = 71.87.$$

If the composition of the binary systems, corresponding to the solubility of substance A at 70.0 °C, is read from the upper half of Fig. 5.21, $x_{AB} = 0.54$ and $x_{AC} = 0.79$, and these points are plotted in the triangular diagram given in Fig. 5.21, then the isotherm for the solubility of A at 70 °C can be constructed by interpolation.

5.21. Example of calculation by the cyclic method.

5.2. Methods based on thermodynamic concepts

The methods treated in Section 5.1 have one disadvantage in common: they are not based on concrete concepts of the behaviour of the system but assume, to a certain extent, additivity of the properties of individual components and attempt to express deviations from additivity by using various geometric constructions. Hence these methods need not necessarily yield correct results. For this reason, much effort has been devoted to attempts to calculate the behaviour of multicomponent systems on the basis of the thermodynamic description of phase equilibria. If a system meets the conditions imposed during derivation of the theoretical relationships, the resulting equations must reliably describe the behaviour of the system. Of course, the theory of concentrated solutions does not permit a priori calculation of the behaviour of the system from the thermodynamic properties of the pure components; however, if a satisfactory equation is obtained from the theory and is modified to express concrete systems by using a number of adjusted parameters, the results thus obtained are still substantially more reliable than results correlated merely on the basis of geometric similarity.

5.2.1. *The method based on the electrolyte solution theory*

For saturated solutions of an electrolyte, which dissociates according to the equation

$$B_bC_c \ldots \rightleftarrows bB + cC + \ldots \tag{5.91}$$

at constant temperature holds

$$a_B^b \cdot a_C^c \ldots = \text{constant} = S_a \tag{5.92}$$

or, if the activities, a, are replaced by the molalities, m, and activity coefficients, γ,

$$m_B^b \cdot m_C^c \ldots \gamma_B^b \cdot \gamma_C^c \ldots = S_a \tag{5.93}$$

The first part of the product on the left-hand side represents the analytical solubility product, S. This property can be considered to be constant only for very poorly soluble substances whose concentrations are very low in saturated solutions and whose activity coefficients thus approach unity. This situation is common in analytical chemistry, where the solubility product finds its principal use. With increasing concentrations of ions in the solution, their activity coefficients progressively differ more from unity and as S_a is a constant, the analytical solubility product varies according to the equation

$$S_a = S\gamma_B^b \gamma_C^c \ldots \tag{5.94}$$

Hence, if it were possible to calculate activity coefficients in solutions from theoretical relationships, it would be possible to determine the electrolyte solubility in multicomponent systems.

The exact relationships derived from the Debye-Hückel theory of electrolyte solutions can be used only for very dilute solutions of substances that contain ions substantially larger than the solvent molecules:

$$\log \gamma_{i\pm} = -z_{i+}z_{i-} \cdot \frac{e^2}{2 \cdot 303} \left(\frac{2\pi N_A \varrho_1}{1000 \varkappa^3 k^3 T^3}\right)^{1/2} \cdot \left(\frac{1}{2} \sum_i C_i z_i^2 \right)^{1/2} \tag{5.95}$$

where $\gamma_{i\pm}$ is the mean activity coefficient of the ith electrolyte, defined by

$$\gamma_{i\pm}^{\nu^i} = \gamma_{i+}^{\nu_{i+}} \gamma_{i-}^{\nu_{i-}} \tag{5.96}$$

where z_{i+} and z_{i-} are the numbers of elemental charges on the cation and anion, respectively, e is one elemental charge, N_A is Avogadro's number, ϱ_1 is the solvent density, $\varkappa$ is the dielectric constant of the solvent, $k = R/N_A$, C_i is the ion molar concentration, ν_{i+} and ν_{i-} are the numbers of cations and anions, respectively, which are formed by the dissociation of the ith electrolyte and $\nu_i = \nu_{i+} + \nu_{i-}$. After substituting the numerical values of the constants, the following relationship is obtained for aqueous solutions at 25 °C

$$\log \gamma_{i\pm} = -0.509 z_{i+} z_{i-} \left(\frac{1}{2} \sum_i C_i z_i^2 \right)^{1/2}. \tag{5.97}$$

As already pointed out, these relationships are valid only for very dilute solutions. Much work has been devoted to attempting to extend the validity of these equations to higher electrolyte concentrations. Firstly, interactions of the particles, which are no longer interpreted as point charges, have been considered; secondly, the effect of a decrease in the solvent dielectric constant due to the action of the ionic field has been taken into account; finally, other more or less empirical corrections have been introduced. The final form of the equation, which holds up to concentrations of ca 4 m, is then

$$\log \gamma_{i\pm} = \frac{-\frac{1}{\nu} \sum \nu_{i\pm} z_{i+} z_{i-} \frac{2.457 \cdot 10^{14}}{R(\varkappa T)^{3/2}} (\sum_i C_i z_i^2)^{1/2}}{1 + A(\sum_i C_i z_i^2)^{1/2}} + Bc + Dc^2 \tag{5.98}$$

where A, B and D are constants. Other theories have been employed to express ion association and have yielded relationships that provide useful results in non-aqueous solvents; however, extrapolation to higher concentrations, when the solute particles are so close together that Van der Waals forces become operative in addition to coulombic forces, can also not be carried out here. Nevertheless, even statistical theories in which both types of forces are considered have not arrived at applicable results.

For the sake of illustration, the basis of an attempt to develop a theory of concentrated electrolyte solutions[1] can be briefly discussed: The ion activity coefficient, γ_i, can be written in the form

$$\gamma_i = \sigma_i \exp (W_i/kT) \tag{5.99}$$

where σ_i is an entropic term expressing changes in the statistical distribution of the particles in the solution compared with that in the pure solvent (water) and W_i is the work required for the formation of the ionic atmosphere around the ith ion. Debye and Hückel neglected the entropy changes during solution formation ($\sigma_i = 1$) and included only coulombic forces in term W_i:

$$W_i = \frac{-(z_i e)^2}{2\varkappa r_0}. \tag{5.100}$$

---

[1] J. Čeleda, Sb. Vys. Sk. Chem.-Technol. (Prague), B 7 (1966) 1.

In highly concentrated solutions a quasi-lattice structure can be assumed, in which the ionic atmosphere is identified with the positions surrounding the central ion and occupied by ions with the opposite charge:

$$z_i e - \int_{r^*}^{r_0} 8\pi c N_A e r^2 \, dr = 0, \tag{5.101}$$

where r_0 is the radius of the ionic atmosphere, r^* is the smallest distance between ions and $\pm 2cN_A e$ is the charge density. If the volume inaccessible to the ion (the molar salting-out volume) is denoted by

$$V_{ij} = \frac{4}{3} \pi N_A r^{*3} \tag{5.102}$$

it then holds that

$$r_0 = r^* \left(1 + \frac{z_i}{2V_{ij}C}\right)^{1/3}. \tag{5.103}$$

By using eqns. 5.99—5.103 and the condition for concentrated solutions it holds that

$$\frac{z_i}{2V_{ij}C} \ll 1, \tag{5.104}$$

$$\log \gamma_i = \log \sigma_i - \frac{A_{ij} z_i^2}{1 + \frac{z_i}{12 V_{ij} C}}, \tag{5.105}$$

where

$$A_{ij} = 1/2 \cdot \frac{e^2}{2.30 r^* \varkappa k T}. \tag{5.106}$$

A substantial portion of the water molecules is replaced in the lattice structure by ions and a number of water molecules are within the reach of force field of the ion. On the other hand, ions are repulsed (salted-out) in the range of their electrostatic field. Of course, the fields of force partially overlap in concentrated solutions. If the salting-out forces do not overlap, then

$$\sigma_i = \left(1 - \frac{2C}{z_j} V_{ij}\right)^{-1} \tag{5.107}$$

as ion i reacts only with oppositely charged ions, j, the molar concentration of which around the ith ion is $2C/z_j$. When salting-out zones overlap, it can be shown that

$$1/\sigma_i = \left(1 - \frac{2C}{z_j} V_{ij}^0\right) \exp\left[-\frac{2C}{z_j}(V_{ij} - V_{ij}^0)\right] \tag{5.108}$$

where V_{ij}^0 is the non-overlapping part of the salting-out zone. For

$$Q_{ij} = \frac{2}{2.30} \cdot \frac{V_{ij}}{z_i z_j} \tag{5.109}$$

it can be shown that

$$\log \gamma_i = \frac{-z_i^2 A_{ij}}{\left(1 + \dfrac{nz_i}{12 V_{ij} \cdot m}\right)^{1/n}} + z_i Q_{ij} m \tag{5.110}$$

where m is the molality, expressed in moles per kilogram of water. The only variable parameter in this equation is the smallest distance between ions, r^*, included in constants A_{ij} and V_{ij}. For the mean activity coefficient of the electrolyte, the relationship becomes

$$\log \gamma_\pm = \frac{-z_+ z_- A}{\left(1 + \dfrac{z_+^2 + z_-^2}{z_+ + z_-} \cdot \dfrac{n}{12 V m}\right)^{1/n}} + \frac{2 z_+ z_-}{z_+ + z_-} \cdot Q m. \tag{5.111}$$

Eqn. 5.111 can be simplified by a suitable choice of n; for example, for $n = 3$ and for

$$V^+ = V \cdot \frac{z_+ + z_-}{z_+^2 + z_-^2} \tag{5.112}$$

the following form is obtained:

$$\frac{1}{m} \log \gamma_\pm = (-z_+ z_- A) \frac{1}{m\left(1 + \dfrac{1}{4 V^+ m}\right)^{1/3}} + \frac{2 z_+ z_-}{z_+ + z_-} Q. \tag{5.113}$$

Hence, by plotting $1/m \log \gamma_\pm$ against $1/[m(1 + 1/4 V^+ m)^{1/3}]$ a straight line is obtained for a correctly chosen V^+, from which the variable parameter r^* can be evaluated.

$$Q'_{ij} = \frac{2 V_{ij} - K_{ij}}{2.30 z_i z_j} \tag{5.114}$$

Note: Data on electrolyte activities can be found in the literature:

a) pure salts in water:

R. A. Robinson and R. H. Stokes, *Electrolyte Solutions*, Butterworths, London, 1959.

b) mixtures of salts:

H. S. Harned, and B. B. Owen, *The Physical Chemistry of Electrolyte Solutions*, Reinhold, New York, 1958.

If the ions are partially associated, the last term on the right-hand side of eqns. 5.110 and 5.111 or 5.113 changes; instead of Q_{ij} (eqn. 5.109), the constant appears, where K_{ij} is the association constant. Characteristic values of these constants and parameters are given in Table 5.3. By using constants A, V and Q, available from tabulated data on binary electrolyte solutions, the activity coefficients of ions in mixtures of any composition can then be calculated. From eqns.

5.110, 5.112 and 5.114, it follows for the activity coefficient of the ith ion in a mixture that

$$\log \gamma_i = -z_i^2 \sum_j \left(\frac{A_{ij} \cdot m_j/m}{\left(1 + \frac{z_i}{4 V_{ij}^\pm m}\right)^{1/3}} \right) + z_i \cdot \sum_j Q'_{ij} m_j \quad (5.115)$$

Table 5.3. Characteristic values of the constants in eqn. 5.110—5.114.

Electrolyte	A_{ij}	V_{ij}^*	Q'_{ij}	V_{ij}	K_{ij}	r^* (10⁻¹⁰ m) 5.106	r^* (10⁻¹⁰ m) 5.102
HCl	0.294	0.167	0.127	0.167	0.03	5.3	4.5
LiCl	0.319	0.208	0.117	0.208	0.15	4.9	4.4
NaCl	0.345	0.167	0.051	0.167	0.22	4.5	4.0
KCl	0.354	0.152	0.019	0.152	0.26	4.4	3.9
RbCl	0.372	0.152	0.015	0.152	0.27	4.2	3.9
CsCl	0.391	0.185	0.007	0.185	0.35	4.0	4.2
LiBr	0.323	0.208	0.138	0.208	0.10	4.8	4.4
NaBr	0.364	0.139	0.072	0.139	0.11	4.3	3.8
KBr	0.336	0.167	0.022	0.167	0.28	4.6	4.0
RbBr	0.358	0.167	0.008	0.167	0.31	4.4	4.0
CsBr	0.397	0.185	0.005	0.185	0.36	3.9	4.2
HI	0.300	0.167	0.188	0.167	0.00	5.2	5.3
NaI	0.348	0.139	0.092	0.138	0.07	4.5	3.8
KI	0.368	0.208	0.024	0.208	0.36	4.2	4.4
RbI	0.377	0.152	0.011	0.152	0.28	4.1	3.9
CsI	0.327	0.278	−0.022	0.278	0.60	4.8	4.8
LiOH	0.348	0.208	0.000	0.208	0.42	4.5	4.4
NaOH	0.386	0.139	0.071	0.139	0.12	4.0	3.8
KOH	0.331	0.167	0.106	0.167	0.09	4.7	4.0
KF	0.360	0.167	0.054	0.167	0.21	4.3	4.0
HNO₃	0.309	0.185	0.076	0.185	0.20	4.3	4.0
MgCl₂	0.333	0.167	0.213	0.28	0.00	4.7	4.8
CaCl₂	0.334	0.185	0.168	0.31	0.18	4.7	5.0
SrCl₂	0.312	0.238	0.125	0.40	0.47	5.0	5.4
BaCl₂	0.303	0.278	0.079	0.46	0.72	5.2	5.7
MgBr₂	0.327	0.167	0.271	0.28	0.00	4.8	4.8
CaBr₂	0.323	0.208	0.208	0.35	0.16	4.8	5.2
SrBr₂	0.323	0.208	0.177	0.35	0.23	4.8	5.2
BaBr₂	0.303	0.238	0.127	0.40	0.47	5.1	5.4
HgI₂	0.316	0.185	0.321	0.31	0.00	4.9	5.0
CaI₂	0.309	0.208	0.256	0.35	0.03	5.0	5.2
SrI₂	0.325	0.185	0.241	0.31	0.00	4.8	5.0
BaI₂	0.277	0.278	0.196	0.46	0.42	5.6	5.7
AlCl₃	0.333	0.208	0.308	0.52	0.10	4.7	5.9
ScCl₃	0.325	0.232	0.260	0.58	0.36	4.8	6.1
YCl₃	0.308	0.278	0.215	0.70	0.73	5.1	6.5
LaCl₃	0.288	0.333	0.165	0.83	1.16	5.4	6.9
CeCl₃	0.298	0.300	0.176	0.75	0.96	5.2	6.7

where the subscript j denotes all of the ions with the opposite charge. According to eqn. 5.115, the effect exerted on the activity coefficient of an arbitrary ion is the sum (on a logarithmic scale) of the effects of all oppositely charged ions, i.e., it is manifested in the additivity of the contributions from its combinations with all oppositely charged ions.

If the activity coefficients of all ions present are known, their values can be substituted into eqn. 5.94 or into its logarithmic form

$$\log S_a = \log S + \sum_i \nu_i \log \gamma_i. \tag{5.116}$$

It should be borne in mind that when the substance crystallizes as a hydrate, the activity coefficient of water must also be considered during the calculation. The $\log S_a$ value is determined as a limiting case, i.e. from the solubility of the pure salt in water.

Example: The solubility of NaCl in the system sodium chloride (1) — potassium chloride (2) — water at 20 °C is to be calculated if the solution contains 100 g of potassium chloride per 1000 g of water. The solubility of sodium chloride in water at this temperature is 360 g sodium chloride per 1000 g of water. Firstly, the activity solubility product, S_a, is calculated from the solubility of sodium chloride in water, m_{01}:

$$m_{01} = 360/M_1 = 6.16 \text{ moles per 1000 g of water}$$

$$S_1^0 = 6.16^2 = 37.95.$$

From eqn. 5.115, it follows, after substitution of the numerical values (taken from table 5.3), that

$$\log \gamma_{Na} = -1^2 \cdot \frac{0.345 \cdot 1}{\sqrt{1 + \frac{1}{4(0.167)(6.16)}}} + 1(0.051)(6.16) = -0.006\,68$$

$$\log \gamma_{Cl} = \log \gamma_{Na} = -0.006\,68.$$

On substitution into eqn. 5.116,

$$\log S_a = 1.5792 - 0.0134 = 1.5658.$$

An analogous calculation for the solution is now performed. For this purpose the value

$$m_2 = \frac{100}{M_2} = 1.34$$

must be calculated and the m_1 value assessed; if the assessment approaches sufficiently closely to the calculated value, the calculation is finished, otherwise the assessment must be corrected and the calculation repeated. A value of $m_1 = 5.5$ is chosen:

$$\log \gamma_{Na} = -1^2 \cdot \frac{0.345 \cdot 1}{\sqrt[2]{1 + \frac{1}{4(0.167)(6.84)}}} + 1(0.051)(6.84) = 0.025\,87$$

The contribution to the chloride ion activity coefficient comes from both sodium and potassium ions:

$$\log \gamma_{Cl} = -1^2 \cdot \frac{(0.345) \cdot \left(\frac{5.5}{6.84}\right)}{\sqrt[3]{1 + \frac{1}{4(0.167)(6.84)}}} + 1(0.051)(5.5) -$$

$$-1^2 \cdot \frac{0.354 \left(\frac{1.34}{6.84}\right)}{\sqrt[3]{1 + \frac{1}{4(0.152)(6.84)}}} + 1(0.019)(1.34) = -0.018\,28$$

On substitution into eqn. 5.116 we obtain

$$\log S = 1.5658 - 0.0076 = 1.5582$$

whence

$$S = 36.15 = m_1(m_1 + 1.34)$$

yielding $m_1 = 5.38$; this value is sufficiently close to the original assessment. Therefore, the composition of the solution is

$$\begin{aligned}
5.38 M_1 &= 314.46 \text{ g NaCl} = 22.23\,\% \\
& 100.00 \text{ g KCl} = 7.07\,\% \\
& 1000.00 \text{ g H}_2\text{O} = 70.70\,\% \\
\text{Total} &= 1414.46 \text{ g}.
\end{aligned}$$

The result can be compared with the experimental values of 6.98 % of potassium chloride and 22.65 % of sodium chloride.

5.2.2. *The theory of conformal ionic solutions*

In this section, the application of the statistical-mechanical theory of conformal ionic solutions to reciprocal salt pairs will be demonstrated. This theory cannot be used for a priori precise computation of the liquidus temperatures; however, it roughly predicts the shape of isotherms in the diagram.[1]

---

[1] M. Blander and L. R. Topol, Inorg. Chem., 5 (1966) 1641.

Let the conversion reaction

$$\text{AC (3)} + \text{BD (2)} \rightleftarrows \text{AD (1)} + \text{BC (4)} \tag{5.117}$$

be considered. For the free excess energy of mixing, ΔG_m^E, of three salts AC, BC and BD, it follows from the conformal ionic solution theory that

$$\Delta G_m^E = x_A x_D \Delta G° + x_D \Delta G_{12}^E + x_C \Delta G_{34}^E + x_A \Delta G_{13}^E + x_B \Delta G_{24}^E + x_A x_B x_C x_D \lambda \tag{5.118}$$

where $\Delta G°$ is the standard molar free energy of reaction 5.117, x_i are the molar fractions of the ions and ΔG_{ij}^E are the free excess energies of mixing for the binary systems:

$$\Delta G_{12}^E = x_A x_B \lambda_{12} \tag{5.119}$$

$$\Delta G_{13}^E = x_C x_D \lambda_{13} \tag{5.120}$$

$$\Delta G_{34}^E = x_A x_B \lambda_{34} \tag{5.121}$$

$$\Delta G_{24}^E = x_C x_D \lambda_{24} \tag{5.122}$$

The λ_{ij} symbols in eqns. 5.119—5.122 are energy parameters, depending solely on the thermodynamic properties of the binary systems. The λ value in eqn. 5.118 cannot be calculated purely theoretically but it is known to be proportional to $(\Delta G°)^2$; an approximation following from the lattice theory is given by the relationship

$$\lambda = -(\Delta G°)^2/2ZRT \tag{5.123}$$

where Z is the lattice coordination number with a value between 4 and 6. The activity coefficient of any component is then given by the equation

$$RT \ln \gamma_i = \frac{\partial n \Delta G_m^E}{\partial n_i} = \frac{\partial n \Delta G_m^E}{\partial n_+} + \frac{\partial n \Delta G_m^E}{\partial n_-}. \tag{5.124}$$

For example, for compound BD(2) the relation

$$RT \ln \gamma_2 = x_A x_C \Delta G° + x_A x_C (x_C - x_D) \lambda_{13} + x_C (x_A x_D + x_B\ x_C) \lambda_{24} + x_A (x_A x_D + x_B x_C) \lambda_{12} + x_A x_C (x_A - x_B) \lambda_{34} + x_A x_C (x_A x_D + x_B x_C - x_B x_D) \lambda \tag{5.125}$$

is obtained, and analogous relationships can be written for the other components. If the dependence of the compound activity coefficients on the composition of the system is known, it is then possible to write, by using modified eqn. 1.57,

$$R \ln a_2 = -\Delta H_{me} \left(\frac{1}{T} - \frac{1}{T_{me}} \right) = R \ln x_B x_D \gamma_2 \tag{5.126}$$

where T_{me} is the melting point of the pure salt and ΔH_{me} is its latent heat of fusion. As this method is not employed for the exact calculation of the behaviour of the system, but for obtaining a general topological description (i.e., the shape of the diagram), it was possible to neglect the temperature dependence of the latent heat of fusion in eqn. 5.126. Hence, if thermodynamic parameters $\Delta G°$, ΔH_{me} and λ_{ij} are known, the course of the liquidus temperatures can be calculated.

The values of parameter λ_{ij} are selected so that they conform as well as possible with the binary eutectic point, i.e.,

$$RT \ln \gamma_i = \lambda_{ij} x_j^2 \qquad (5.127)$$

$$RT \ln \gamma_j = \lambda_{ij} x_i^2. \qquad (5.128)$$

Combination of eqns. 5.126 — 5.128 leads to relationships for the eutectic temperature, T_E:

$$T_E = (\lambda_{ij}(1-x_i)^2 + \Delta H_i)/[(\Delta H_i/T_{i\,\text{me}}) - R \ln x_i] \qquad (5.129)$$

$$T_E = (\lambda_{ij} x_i^2 + \Delta H_j)/[(\Delta H_j/T_{j\,\text{me}}) - R \ln (1-x_i)] \qquad (5.130)$$

from which the λ_{ij} value can be calculated from the known eutectic temperature and eutectic composition. During the calculation T_E is employed as an experimentally determined constant and the x_i and λ_{ij} values are variables, for which eqns. 5.129 and 5.130 can be solved. The equation for the liquidus temperature is then obtained by combining eqns. 5.125 and 5.126:

$$T[(\Delta H_{BD}/T_{\text{me}}) - R \ln x_B x_D]$$
$$= \Delta H_{BD} + x_A x_C [\Delta G^\circ + (x_C - x_D) \lambda_{13} + \lambda (x_A - x_B) \lambda_{34}] +$$
$$+ x_C (x_A x_D + x_B x_C) \lambda_{24} + x_A (x_A x_D + x_B x_C) \lambda_{12} +$$
$$+ x_A x_C (x_A x_D + x_B x_C - x_B x_D) \lambda. \qquad (5.131)$$

Example: The shape of the liquidus isotherms was calculated for a system of fused salts, Li, K || F, Cl. The following data were employed in the calculation (Li $\equiv$ B, F $\equiv$ D):

System	Measured eutectic temperature	composition	λ_{ij}	Calculated eutectic composition
LiF — KF	492	49.5 % LiF	−17 781	50.8 % LiF
LiF — LiCl	498	30.0 % LiF	−1528	30.0 % LiF
KCl — LiCl	354	57.5 % LiCl	−17 304	58.8 % LiCl
KCl — KF	606	45.0 % KF	−408	43.2 % KF

Substance	T_{me}	ΔH_{me}
LiF	1121	27 089
KF	1129	28 261
LiCl	878	19 929
KCl	1045	26 544

$Z = 6$, $\Delta G^\circ = 68.2$ kJ/mol.

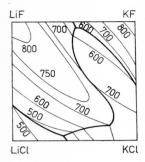

Using these data, the liquidus temperatures given by eqn. 5.131 were evaluated using a computer. The experimental (upper) and calculated (lower) phase diagrams are compared in Fig. 5.22. It is evident that the calculation can yield sufficiently precise data for rapid information about a given reciprocal salt pair system. The method can, of course, also be applied to other types of systems.

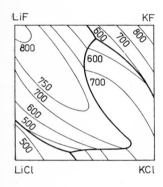

5.22. Example of calculation using the theory of conformal solutions.

5.2.3. The Zdanovskii method and modifications

The basis of the Zdanovskii method[1] is the postulate that the solubility of a salt in solutions of various electrolytes with a common ions is additive when the final solution is prepared by mixing solutions of salts with the same water activity. On the basis of knowledge of the solubility in two- and three-component systems and water activity data, the solubility of the salt in a multicomponent system system can be calculated.

According to eqns. 1.46 and 1.47, characterizing phase equilibrium by equality of the chemical potentials of the substance in the solid and liquid phases, the relationship

$$\mu_s^0 + RT \ln f_s = \mu_l^0 + RT \ln m + RT \ln \gamma \qquad (5.132)$$

can be written, where μ_s^0 and μ_l^0 are the standard chemical potentials of the substance in the solid and liquid phases, respectively, f_s is its fugacity in the solid phase, m its molality and γ its activity coefficient defined in terms of the molality:

$$a_i = m_i \gamma_i. \qquad (5.133)$$

[1] A. B. Zdanovskii, Zh. Fiz. Khim., 22 (1948) 1475 and 1486.

The Gibbs-Duhem equation is valid for the activities of the individual components in equilibrium

$$\sum_{i=1}^{s} x_i \, d \ln a_i = 0 \qquad [T, P, \text{equil}]. \tag{5.134}$$

The calculation can begin with two three-component systems, as follows. Solution I contains x'_A of salt A with activity a'_A, x'_B of salt B with activity a'_B and x'_D of water with activity a'_D. Solution II contains x''_A of salt A with activity a''_A, x''_C of salt C with activity a''_C and x''_D of water with activity a''_D. The composition of the solutions is such that the activities of substance A and water are the same in both solutions:

$$a'_A = a''_A = a_A \tag{5.135}$$

$$a'_D = a''_D = a_D. \tag{5.136}$$

It must also be valid that

$$x'_A + x'_B + x'_D = 1. \tag{5.137}$$

$$x''_A + x''_C + x''_D = 1. \tag{5.138}$$

Solutions I and II are then mixed in a molar ratio of $x : (1-x)$ so that the overall number of moles in the mixture remains equal to unity. It follows from eqn. 5.134 that

$$[xx'_A + (1-x) x''_A] \, d \ln a_A + xx'_B \, d \ln a_B +$$
$$+ (1-x) x''_C \, d \ln a_C + [xx'_D + (1-x) x''_D] \, d \ln a_D = 0 \tag{5.139}$$

and, as eqns. 5.135 and 5.136 are simultaneously valid, this is reduced to give

$$[xx'_A + (1-x) x''_A] \, d \ln a_A = 0 \tag{5.140}$$

$$[xx'_D + (1-x) x''_D] \, d \ln a_D = 0. \tag{5.141}$$

Eqn. 5.139 simplifies to give

$$xx'_B \, d \ln a_B + (1-x) x''_C \, d \ln a_C = 0. \tag{5.142}$$

In eqn. 5.142, $a_B \neq a'_B$ and $a_C \neq a''_C$, and therefore only additivity of the isoactive solutions is considered.

A number of ternary solutions, 1, 2, 3, ..., with compositions $x_1 + y_1$, $x_2 + z_2$ ..., etc., can now be compared; x is the content of the component whose solubility is to be calculated and, y, z, etc., are the contents of the other electrolytes. All of the solutions have the same constant water activity. If these solutions are mixed in a ratio such that α_i is the weight fraction of the ith solution in the mixture, the concentrations of the individual components in the resulting solution will be given by the relationships

$$X = \sum_i \alpha_i x_i \tag{5.143}$$

$$Y = \alpha_1 y_1 \tag{5.144}$$

$$Z = \alpha_2 z_2 \tag{5.145}$$

etc. By combining eqns. 5.143 — 5.145, the relationship

$$X = \frac{Y}{y_1} x_1 + \frac{Z}{z_2} x_2 + \ldots \qquad (5.146)$$

is obtained. As already mentioned, the calculation of the water activity in a solution of several electrolytes is based on the assumption that the water activity is constant during the mixing of isopiestic electrolyte solutions. If binary solutions with identical water activities, containing $x_0, y_0, z_0, \ldots$ of salts (in wt.- % or g/l) are mixed in the ratio of weight fractions α_i, the resultant concentrations are given by

$$X = \alpha_1 x_0 \qquad (5.147)$$
$$Y = \alpha_2 y_0 \qquad (5.148)$$
$$Z = \alpha_3 z_0 \qquad (5.149)$$

etc. From these equations, it follows that

$$\frac{X}{x_0} + \frac{Y}{y_0} + \frac{Z}{z_0} + \ldots = 1 \qquad (5.150)$$

or

$$X = x_0 \left(1 - \frac{Y}{y_0} - \frac{Z}{z_0} - \ldots \right). \qquad (5.151)$$

The solubility value to be determined in a multicomponent system is found by solving eqns. 5.146 and 5.151. The most suitable procedure for the solution involves plotting the X values, calculated from eqn. 5.146, and the X values, calculated from eqn. 5.151, against the water activity, a; the intercept of the two curves gives the solution.

Eqns. 5.143 — 5.145 for a ternary system are simplified to the form

$$X = x_1 + (1 - \alpha) x_2 \qquad (5.152)$$
$$Y = \alpha y_1 \qquad (5.153)$$
$$Z = (1 - \alpha) z_2. \qquad (5.154)$$

When constructing constant activity lines, the relationship

$$\Sigma_0 = X + Y + Z = \alpha x_1 + (1 - \alpha) x_2 + \alpha y_1 + (1 - \alpha) z_2 = \text{constant} \qquad (5.155)$$

can be employed, from which the mixing ratio, α, is determined:

$$\alpha = \frac{\Sigma_0 - x_2 - z_2}{x_1 + y_1 - x_2 - z_2}. \qquad (5.156)$$

The mutual solubility of the salts is determined by solution of the following equations (analogous to eqns. 5.146 and 5.151).

Solution saturated with salt I

$$X = \frac{Y}{y_1'} x_1' + \frac{Z}{z_2'} x_2' + \ldots \qquad (5.157)$$

$$Y = y_1' \left(1 - \frac{Z}{z_2'} - \ldots \right) \qquad (5.158)$$

Solution saturated with salt II:

$$X = x_1'' \left(1 - \frac{Z}{z_2''} - \ldots \right) \qquad (5.159)$$

$$Y = \frac{X}{x_1''} y_1'' + \frac{Z}{z_2''} y_2'' + \ldots \qquad (5.160)$$

Various water activity values are again chosen and the intercept of the two curves in coordinates X versus Y is determined.

Example: Calculation of the water activity in a solution saturated with sodium chloride and potassium chloride at 25 °C. The solution contains 20.42 % of sodium chloride and 11.14 % of potassium chloride. From eqn. 5.150, it follows that

$$\frac{20.42}{x_0} + \frac{11.14}{y_0} = 1$$

The x_0 and y_0 values are found from the graph of the dependence of the water activity on the composition of the corresponding binary solutions: a water activity close to the assumed value, e.g. $a = 0.73$, is chosen and concentrations $x_0 = 27.8$ and $y_0 = 37.8$ are found for this value. On substitution into the above equation, a value of unity is not obtained, but

$$\frac{20.42}{27.8} + \frac{11.14}{37.8} = 1.0292.$$

The x_0 and y_0 values are therefore corrected:

$$x_0 = 1.0292 \cdot 27.8 = 28.61 \% \pm \text{of sodium chloride}$$
$$y_0 = 1.0292 \cdot 37.8 = 38.71 \% \pm \text{of potassium chloride}$$

and the corresponding water activities are again found for these values in the appropriate binary graphs: $a = 0.716$, $a = 0.720$. The mixing ratio for the two solutions is calculated from eqns. 5.147 and 5.148

$$\alpha_1 = 20.42 / 28.61 = 0.714; \quad \alpha_2 = 11.14 / 38.71 = 0.288.$$

Thus the water activity is

$$a = 0.714 \cdot 0.716 + 0.288 \cdot 0.720 = 0.718.$$

Calculation of the solubility in a ternary system can also be performed graphically, as is illustrated in the anhydrous projection given in Fig. 5.23. An example of the construction of points on the curve of solutions saturated with substances A and B can be given. The first point, E, is obtained by transferring the eutectic point to traingle side $\overline{AB}$. The second (third) point is constructed as follows.

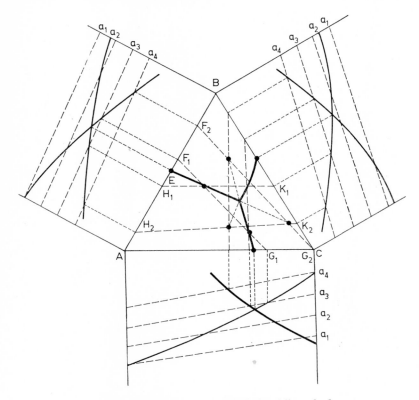

5.23. Calculation using the Zdanovskii method.

The intercepts of the solubility isotherm for substance A with the constant water activity line $a_3(a_4)$ in ternary mixtures A + B + water and A + C + water are transferred to the appropriate triangle sides, giving points $F_1(F_2)$ and $G_1(G_2)$. In an analogous manner, the images of the intercepts of the solubility isotherm for substance B with the constant water activity line $a_3(a_4)$ are obtained and denoted as points $H_1(H_2)$ and $K_1(K_2)$. The intercept of connecting lines $\overline{F_1G_1}$ and $\overline{H_1K_1}$ ($\overline{F_2G_2}$ and $\overline{H_2K_2}$) is the second (third) point lying on the connecting line of the eutonic points of substances A and B. The remaining two curves are constructed in an analogous manner; all three curves intersect in a single point — the quaternary eutectic point.

Of course, it may be difficult to obtain data on the water activity in ternary, or at least binary, systems. Zdanovskii then recommends that the calculation should be carried out for isohydric solutions (the number of moles of water per 100 g — equiv. of electrolyte is constant) instead of isopiestic solutions (a = constant). The method thus modified is less precise, but can often yield satisfactory data.

Example: The solubility is to be determined in a sodium chloride-potassium chloride — hydrogen chloride-water system at 25 °C. The solubilities in the partial ternary systems are given by the equations

$$\log x_{NaCl} = 1.42292 - 0.0354\,(1 - a)\,y_{KCl}$$
$$\log x_{NaCl} = 1.42292 - 0.1070\,(1 - a)\,y_{HCl}$$
$$\log x_{KCl} = 1.42357 - 0.0283\,(1.358 - a)\,y_{NaCl}$$
$$\log x_{KCl} = 1.42357 - 0.0402\,(1.654 - a)\,y_{HCl}.$$

The composition that corresponds to the eutonic point, i.e., to a solution saturated with potassium chloride and sodium chloride in a solution containing 8.61 % of hydrogen chloride, can then be calculated. Using the above equations, it can be calculated that a content of 8.61 % of hydrogen chloride in the ternary system corresponds to a sodium chloride solubility (in the absence of potassium chloride) of 14 % and to a potassium chloride solubility (in the absence of sodium chloride) of 13 %. Now, the water activity is assessed from the graph of $x - y$ versus a (a = 0.66), and the composition of ternary solutions saturated with sodium chloride is calculated from the appropriate relationships:

A: 9.35 % NaCl + 12.45 % HCl
B: 12.7 % NaCl + 26.3 % KCl

In an analogous manner, the composition of the ternary solutions saturated with potassium chloride is obtained:

C: 6.75 % KCl + 14.85 % HCl
D: 7.95 % KCl + 26.0 % NaCl.

The pairs of solutions are then mixed in ratios so that the required hydrogen chloride content in the mixture is obtained: For saturated sodium chloride solutions, solutions A and B are mixed in the ratio 8.61 / 12.45: (12.45 — 8.61)/ 12.45, so that they contain

I: sodium chloride = $9.35\,\dfrac{8.61}{12.45} + 12.7\,\dfrac{12.45 - 8.61}{12.45} = 10.33$ %

potassium chloride = $26.3\,\dfrac{12.45 - 8.61}{12.45} = 8.11$ %

hydrogen chloride = 8.61 %.

Solutions C and D, saturated with potassium chloride, are mixed in the ratio
8.61 / 14.85 : (14.85 — 8.61) / 14.85, so that they contain

II: sodium chloride $= 26.0 \dfrac{14.85 - 8.61}{14.85} = 10.93 \%$

potassium chloride $= 6.725 \dfrac{8.61}{14.85} + 7.95 \dfrac{14.85 - 8.61}{14.85} = 7.25 \%$

hydrogen chloride $= 8.61 \%$.

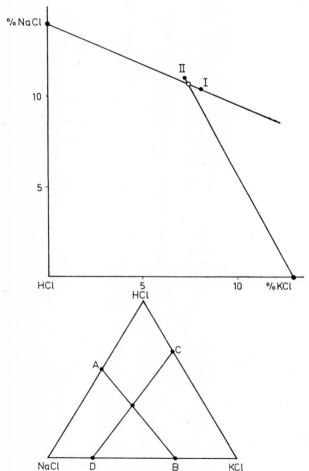

5.24. Example of caluclation by the Zdanovskii method.

A graph is now constructed (Fig. 5.24 top), in which the sodium chloride solubility is plotted on the ordinate and the potassium chloride solubility on the abscissa. Two points are obtained from the solubility data in the three-component systems described above and two further points are given by solutions I and II. The lines connecting the points of the sodium chloride and potassium chloride solubilities, which could be made more precise by calculation of more points, intersect at a point corresponding to the composition to be determined: 10.7 % of sodium chloride, 7.4 % of potassium chloride and 8.61 % of hydrogen chloride. After re-calculation to (sodium chloride + potassium chloride + hydrogen chloride) = 100 %, this point can be represented as the figurative point of the system in the triangular diagram (Fig. 5.24 bottom). The figurative point (without knowledge of the water content) can, of course, be obtained very rapidly graphically: if the composition of solutions A, B, C and D is re-calculated to values $x + y = 100$ %, we then obtain

Point	% NaCl	% KCl	% HCl
A	43	—	57
B	32.5	67.5	—
C	—	31.2	68.8
D	76.5	23.5	—

These points are represented in the triangular diagram and the intercept of connecting lines $\overline{AB}$ and $\overline{CD}$ gives the figurative point, identical with the previous result.

5.2.4. *Expansion of relative activity coefficients*[1,2,3]

A three-component system consisting of a solvent (0) and two further components (1 and 2) can be considered. According to eqns. 1.47 and 1.48, the phase equilibrium between the solid (') and liquid (") phases is characterized by equality of the chemical potentials of a given component in the two phases:

$$\mu'_1 = \mu''_1 \tag{5.161}$$

$$\mu'_2 = \mu''_2. \tag{5.162}$$

According to eqn. 1.46, relationships

$$\mu'_i = \mu_i^{0'} + RT \ln (\gamma'_i y_i) \tag{5.163}$$

and

$$\mu''_i = \mu_i^{0''} + RT \ln (\gamma''_i m_i) \tag{5.164}$$

hold for the chemical potentials, where y_i is the molar fraction of the ith component in the solid phase, m_i is its molality in the solution and γ'_i or γ''_i its activity coefficients.

[1] E. Erdös, Chem. Listy, *51* (1957), 1632.
[2] J. Nývlt, Collect. Czech. Chem. Commun., *34* (1969), 2348.
[3] J. Nývlt, A. Majrich and H. Kočová: Collect Czech. Chem. Commun., *35* (1970) 165.

It follows from eqns. 5.161 and 5.164 for the phase equilibrium condition that
$$\ln (m_i \gamma_i'') = \ln (y_i \gamma_i' \lambda_i) \tag{5.165}$$
where
$$\ln \lambda_i = \frac{\mu_i^{0'} - \mu_i^{0''}}{RT}. \tag{5.166}$$

If a similar consideration holds for binary systems, i.e., for the solubility of the i-th component in the pure solvent, the simplifying conditions
$$\gamma_{0i}' = 1 \tag{5.167}$$
and
$$y_{0i} = 1 \tag{5.168}$$
are valid. Then the relationship
$$\ln(m_{0i} \gamma_{0i}'') = \ln \lambda_i \tag{5.169}$$
is obtained, giving, in combination with eqn. 5.165,
$$\log \frac{m_i}{m_{0i} y_i} = \log \frac{\gamma_i' \gamma_{0i}''}{\gamma_i''}, \tag{5.170}$$

If the symbols for the relative molality
$$x_i = \frac{m_i}{m_{0i}} \tag{5.171}$$
and for the relative activity coefficient
$$\xi_i = \frac{\gamma_i''}{\gamma_{0i}''} \tag{5.172}$$
are introduced, eqn. 5.170 assumes the basic form
$$\log \frac{x_i}{y_i} = -\log \frac{\xi_i}{\gamma_i'} = \varphi_i \qquad [T, P, \text{sat}]. \tag{5.173}$$

This equation can readily by rearranged for a three-component system to give
$$\log \frac{x_1 y_2}{x_2 y_1} = \varphi_1 - \varphi_2. \tag{5.174}$$

This equation will be encountered in the next chapter as the Dörner—Hoskins distribution law.

Eqn. (5.173) holds when component 1 is not dissociated in solution. If this component is an electrolyte of the $M_{\nu_+} X_{\nu_-}$ type, dissociating into M^{z+} and X^{z-} ions, according to the scheme
$$M_{\nu_+} X_{\nu_-} \rightleftarrows \nu_+ M^{z+} + \nu_- X^{z-} \tag{5.175}$$
its chemical potential is given by the equation
$$\mu_1 = \nu_{1+} \mu_{1+} + \nu_{1-} \mu_{1-}. \tag{5.176}$$

In an analogous manner, the mean activity coefficient is given by

$$\gamma_1^{\nu_1} = \gamma_{1+}^{\nu_{1+}} \gamma_{1-}^{\nu_{1-}} \tag{5.177}$$

where

$$\nu_1 = \nu_{1+} + \nu_{1-}. \tag{5.178}$$

Then eqn. 5.173 assumes the form

$$\log \frac{x_{i+}^{\nu_{i+}/\nu_i} \cdot x_{i-}^{\nu_{i-}/\nu_i}}{y_{i+}^{\nu_{i+}/\nu_i} \cdot y_{i-}^{\nu_{i-}/\nu_i}} = \varphi_i \tag{5.179}$$

If both components 1 and 2 are ionic substances with a common cation, the folowing balance is valid

$$x_{i+} = \nu_{i+} x_1 + \nu_{i+} x_2 \frac{m_{20}}{m_{10}} \tag{5.180}$$

$$x_{i-} = \nu_{i-} x_i \tag{5.181}$$

$$y_{i+} = \frac{\nu_{i+}}{\nu_i} y_i + \frac{\nu_{i-}}{\nu_i} y_2 \tag{5.182}$$

$$y_{i-} = \frac{\nu_{i-}}{\nu_i} y_i. \tag{5.183}$$

As solid solutions are formed only by salts of similar types, it can be assumed that

$$\nu_{1+} = \nu_{2+}. \tag{5.184}$$

Then, by substitution of eqns. 5.180—5.184 into the basic equation, (eqn 5.179), and by rearrangement the equation

$$\Phi_i = \varphi_i + \frac{1}{\nu_i} \log \frac{\nu_i^{\nu_{i+}} + \nu_i^{\nu_{i-}}}{\nu_i^{\nu_i}} y_i^{\nu_{i-}} = \frac{1}{\nu_i} \log \left[x_i^{\nu_{i-}} \left(x_i + x_j \frac{m_{j0}}{m_{i0}} \right)^{\nu_{i+}} \right] \tag{5.185}$$

is obtained. If the two electrolytes have a common anion, then, in an analogous manner:

$$\Phi_i = \varphi_i + \frac{1}{\nu_i} \log \frac{\nu_i^{\nu_{i+}} + \nu_i^{\nu_{i-}}}{\nu_i^{\nu_i}} y_i^{\nu_{i+}} = \frac{1}{\nu_i} \log \left[x_i^{\nu_{i+}} \cdot \left(x_i + x_j \frac{m_{j0}}{m_{i0}} \right)^{\nu_{i-}} \right]. \tag{5.186}$$

Substantially simpler relationships are valid when the components are immiscible in the solid phase. Then,

$$y_1 = 1 \tag{5.187}$$
$$y_2 = 0 \tag{5.188}$$
$$\gamma'_i = 1 \tag{5.189}$$
$$\lambda_1 = 1 \tag{5.190}$$

and the basic equation (eqn. 5.173) assumes the simple form

$$\log x_1 = -\log \xi_1 = \varphi_1 \qquad [T, P, \text{sat}]. \tag{5.191}$$

If compound 1 is dissociated and ions M and X in solution do not originate from any other component, eqn. 5.191 can be rewritten to give

$$\log \left(\frac{S_1}{S_{10}}\right)^{1/\nu_1} = \log x_1 = \varphi_1 \qquad (5.192)$$

where S_{10} and S_1 are the analytical solubility products of component 1 in the pure solvent and in the multicomponent system, respectively. Eqn. 5.191 can be compared with eqn. 5.173. If the presence in solution of two electrolytes with a common cation is assumed, then, by using relationships 5.180—5.183, the equation

$$\log \left[x_1^{\nu_{1-}/\nu_1} \left(\sum_{i=1}^{n} \frac{\nu_{i+} m_{i0}}{\nu_{1+} m_{10}} x_i \right)^{\nu_{1+}/\nu_1} \right] = \varphi_1 \qquad (5.193)$$

is obtained. Similarly, for two electrolytes with a common anion, the equation

$$\log \left[x_1^{\nu_{1+}/\nu_1} \left(\sum_{i=1}^{n} \frac{\nu_{i-} m_{i0}}{\nu_{1-} m_{10}} x_i \right)^{\nu_{1-}/\nu_1} \right] = \varphi_1 \qquad (5.194)$$

is obtained. For all possible combinations of mono — and divalent ions, a common equation is obtained

$$\varphi_1 = \frac{1}{\alpha + \beta} \log x_1^{\alpha}(x_1 + FB)^{\beta}] \qquad (5.195)$$

where

$$B = x_2 \cdot \frac{m_{20}}{m_{10}} \qquad (5.196)$$

$$F = \frac{\nu_{2\mp}}{\nu_{1\mp}} \qquad (5.197)$$

$$\alpha = \nu_{1\pm} \qquad (5.198)$$

$$\beta = \nu_{1\mp}. \qquad (5.199)$$

In eqns. 5.197—5.199, the upper sign is valid for a common anion and the lower for electrolytes with a common cation.

The α, β and F values are given in Table 5.4 for various combinations of ions.

On the right-hand side of eqn. 5.173 or 5.191 there is, of course, still an unknown function involving the relative activity coefficient. Its value must, in general, depend on the composition, temperature and pressure of the system:

$$\varphi_1 = \varphi_1(m_1, m_2, \ldots m_n, T, P). \qquad (5.200)$$

As, however, the composition of the condensed, two-phase system is unambiguously determined by $(N-1)$ concentration values, this general functional dependence can be rewritten in the form

$$\varphi_1 = \varphi_1(m_2, m_3, \ldots m_n), \; [T, P]. \qquad (5.201)$$

Table 5.4. Values of constants α, β and F in eqn. 5.195. Electrolyte denoted as $z_+ - z_-$, common ion bold

Electrolyte 1	Electrolyte 2	p	q	F
1 − 1	**1** − 1	1	1	1
1 − **1**	1 − **1**	1	1	1
1 − 1	**1** − 2	1	1	2
1 − **1**	2 − **1**	1	1	2
1 − 2	**1** − 1	1	2	0.5
1 − 2	**1** − 2	1	2	1
1 − **2**	1 − **2**	2	1	1
1 − **2**	2 − **2**	2	1	1
2 − 1	**1** − 1	1	2	0.5
2 − 1	**2** − 1	2	1	1
2 − 1	**2** − 2	2	1	1
2 − **1**	2 − **1**	1	2	1
2 − 2	**2** − 1	1	1	1
2 − 2	**2** − 2	1	1	1
2 − **2**	1 − **2**	1	1	1
2 − **2**	2 − **2**	1	1	1
no common ions		1	1	0

This form has the advantage of not containing concentration value m_1 and thus permits the explicit expression of x_1 from the basic equation. The expansion of the general function 5.201 into the MacLaurin series with respect to molalities $m_{i \neq 1}$ yields the equation

$$\varphi_1 = \sum_{i=2}^{n} Q_{1i} m_i + \sum_{i=2}^{n} \sum_{j=2}^{n} Q_{1ij} m_i m_j + \ldots \tag{5.202}$$

The number of terms required for the expression of a real system can best be determined graphically: with a three-component system, φ_1/m_2 is plotted against m_2. If a single term suffices for the expression of function φ_1, i.e.,

$$\varphi_1 = Q_{12} m_2 \tag{5.203}$$

or

$$\varphi_1/m_2 = Q_{12} \tag{5.204}$$

then the plot of the dependence will be a horizontal straight line with coordinate Q_{12}. If function φ_1/m_2 is expressed by two terms, i.e.,

$$\varphi_1 = Q_{12} m_2 + Q_{122} m_2^2 \tag{5.205}$$

or

$$\varphi_1/m_2 = Q_{12} + Q_{122} m_2 \tag{5.206}$$

then eqn. 5.206 will be represented by a sloping straight line with slope Q_{122} and an intercept of Q_{12} on the φ_1/m_2 axis. If a curve is obtained instead of a straight line, a greater number of terms for expressing function φ_1 is required. The resulting equation expressing the dependence of the solubility of substance 1 in a multi-component system, is obtained by combining eqn. 5.173, 5.185, 5.186 or 5.195 with expansion 5.202.

The adjustable interaction constants, Q, can be evaluated from the experimental data for three-component systems; these constants can then be employed for concentration or temperature interpolations and also for the calculation of phase equilibria in multicomponent systems.

For example, the solubility of component 1 in a four-component system can be written in terms of three-component systems (the salts are, for example, of the $1 - 1$ type), as follows.

Three-component systems:

$$\frac{1}{2}\log\left[x_1\left(x_1 + \frac{m_{20}}{m_{10}}x_2\right)\right] = Q_{12}m_2 + Q_{122}m_2^2$$

$$\frac{1}{2}\log\left[x_1\left(x_1 + \frac{m_{30}}{m_{10}}x_3\right)\right] = Q_{13}m_3 + Q_{133}m_3^2.$$

Four-component system:

$$\frac{1}{2}\log\left[x_1\left(x_1 + \frac{m_{20}}{m_{10}}x_2 + \frac{m_{30}}{m_{10}}x_3\right)\right] = Q_{12}m_2 + Q_{122}m_2^2 + Q_{13}m_3 + Q_{133}m_3^2 + Q_{123}m_2m_3$$

Constant Q_{123} can be evaluated by using a single experimental value for the four-component system or assessed by using some of the combination rules,[1] for example

$$Q_{123} = (Q_{12} + Q_{13})^{1/2}.$$

Moreover, the constants Q usually depend very little on temperature, as the relative molalities, related to the solubility of the substance in the pure solvent, are employed; hence calculations for other temperatures can also be carried out with the given constants. Finally, it can be seen, as actually follows from comparison of eqn. 5.98 or 5.111 with eqn. 5.202, that the value of Q may by characteristic of the individual ion and therefore its value could be predicted in analogy.

As the calculation is time consuming, computer programs for the evaluation of Q from experimental data and for the back-calculation of solubilities in ternary systems from the corresponding Q constants are described below. The programs are in FORTRAN IV G language and the block schemes are given in Figs. 5.25 and 5.26.

[1] E. Erdös:, Chemické listy, 51, (1957) 1632

FORTRAN IVG program for evaluation of Q

Symbols:

$A(1)$ ...	solubility of substance 1, in grams per 100 g of solvent
$A(2)$ ...	concentration of substance 2, in grams per 100 g of solvent
$P(1)$ ...	solubility of substance 1, in wt. — %
$P(2)$ ...	concentration of substance 2, in wt. — %
$C(1)$ ...	molality of substance 1
$C(2)$ ...	molality of substance 2
AB ...	solubility of substance 1 in the pure solvent, in grams per 100 g of solvent
PB ...	solubility of substance 1 in pure solvent, in wt. — %
CB ...	molality of substance 1 in pure solvent
$M(1)$ ...	molecular weight of substance 1
$M(2)$ ...	molecular weight of substance 2
ALPHA ...	α, see eqn. 5.198 or Table 5.4
BETA ...	β, see eqn. 5.199 or Table 5.4
F ...	see eqn. 5.197 or Table 5.4
$X(1)$ ...	relative molality of substance 1
Y ...	an auxiliary quantity
Z ...	an auxiliary quantity
N ...	an auxiliary quantity
$Q(1)$ ...	Q_{12}
$Q(2)$ ...	Q_{122}

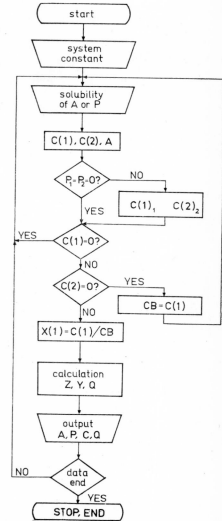

5.25. Block scheme of solubility evaluation.

Punching the input data cards:

	1	10	20	30	40	45
Card No. 1	M (1)	M (2)	ALPHA	BETA	F	
Card No. 2 or	AB		PB			
Other cards or	A (1)	A (2)				
			P (1)	P (2)		

```
C           EVALUATION OF THE SOLUBILITY DATA OF SUBSTANCE (1)
            REAL * 4 A(2), P(2), M(2), ALPHA, BETA, C(2), F, Y, Q, CB, Z, X(1)
            READ (5, 1000) M(1), M(2), ALPHA, BETA, F
    1000    FORMAT (2F10.2, 2F10.0,F5.2)
            WRITE (6, 1001) M(1), M(2), ALPHA, BETA, F
    1001    FORMAT(1H1,40HSOLUBILITY OF (1) IN A TERNARY SYSTEM /5HM(1)=,
           1 F6.2,6H,M(2)=,F6.2,7H,ALPHA=,F3.0,6H,BETA=,F3.0,F=,F3.1 //
           2 57H      A(1)         A(2)          P(1)         P(2)         C(2)        Q        //)
      1     READ (5,1002,END=12) A(1),A(2),P(1),P(2)
    1002    FORMAT (4F10.4)
C           CALCULATION OF RELATIVE MOLALITIES
      2     C(1)=10.*A(1)/M(1)
            C(2)=10.*A(2)/M(2)
            IF (P(1).EQ.0.0.AND.P(2).EQ.0.0) GO TO 3
            C(1)=1000.*P(1)/M(1)*(100.-P(1)-P(2)))
            C(2)=1000.*P(2)/(M(2)*(100.-P(1)-P(2)))
      3     IF(C(1).EQ.0.0) GO TO 1
            IF(C(2).EQ.0.0) GO TO 4
            GO TO 5
      4     CB = C(1)
            GO TO 1
      5     X(1) = C(1)/CB
C           CALCULATION OF THE RELATIVE ACTIVITY COEFFICIENTS
            Z = (X(1)** ALPHA)*((X(1) + F* C(2)/CB) ** BETA)
            Y = (1/ALPHA + BETA))* ALOG10 (Z)
C           CALCULATION OF THE INTERACTION CONSTANTS
            Q = Y/C(2)
      6     WRITE (6,1003) A(1), A(2),P(1),P(2),C(2),Q
    1003    FORMAT (1H, 5F10.4,E10.4)
C           PLOT VALUES OF Q VS. C(2) AND EVALUATE Q(1) AND Q(2)
            GO TO 1
     12     STOP
            END
```

The resulting data are printed in a table with the heading:

SOLUBILITY OF (1) IN A TERNARY SYSTEM

M(1) =	M(2) =	α =	β =	F =	
A(1)	A(2)	P(1)	P(2)	C(2)	Q
...	...	...	...	...	...

The value of $Q = \varphi_1/m_2$ is plotted against $C(2) = m_2$ and the constants Q_{12}, Q_{122}, ... are evaluated graphically.

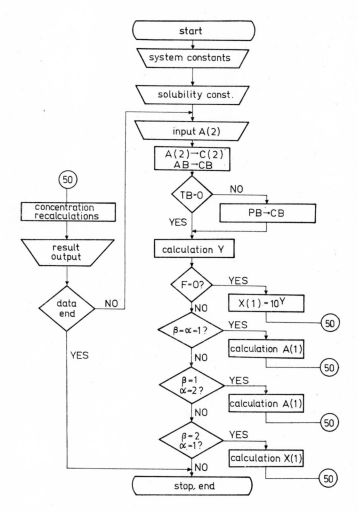

5.26. Block scheme of reversed solubility calculation.

FORTRAN-IV G program for reverse solubility calculation.

Punching of the input data cards:

	1	10	20	30	40	45
card No. 1	M(1)	M(2)	ALPHA	BETA	F	
card No. 2	AB		Q(1)	Q(2)		
or		PB	Q(1)	Q(2)		
Other cards	A(2)					

```
C          CALCULATION OF THE SOLUBILITY OF (1) IN A TERNARY SYSTEM
           REAL*4 M(2),A(2),P(2),C(2),CB,ALPHA,BETA,F,Y,Z,Q(2),AB,X(1),N,PB
           READ (5,1000) M(1), M(2), ALPHA, BETA,F
    1000   FORMAT (2F10.2,2F10.0,F5.2)
           READ (5,1001) AB, PB,Q(1),Q(2)
    1001   FORMAT(3F10.4,E10.4)
           WRITE (6,1002) M(1),M(2),ALPHA,BETA,F,Q(1),Q(2),AB,PB
    1002   FORMAT (10H1,40HSOLUBILITY OF (1) IN THE TERNARY SYSTEM/6H
          1 M(1)=,
          2 F6.2,6H,M(2)=,F6.2,7H,ALPHA=,F3.0,6H,BETA=,F3.0,3H,F=,F3.1,
          3 6H,Q(1)=,F10.4/5H,Q(2)=,E10.4,4H,AB=,F10.4,7H,OR PB=,F10.4//
          4 70H    P(2)       A(2)       C(2)       P(1)       A(1)       C(1)
          5 X(1)         //)
       1   READ (5,1003,END=500) A(2)
    1003   FORMAT (F10.4)
C          CALCULATION OF MOLALITIES
           C(2) = 10.*A(2)M(2)
           CB = 10.* AB/M(1)
           IF (PB.EQ.0.0) GO TO 2
           CB= 1000.* PB/(M(1)*(100.-PB))
       2   Y = Q(1)*C(2) + Q(2)*C(2)** 2.
C          CALCULATION OF X(1)
           IF(F.EQ.0.0) GO TO 10
           IF(BETA.EQ.1.0.AND.ALPHA.EQ.1.0) GO TO 20
           IF(BETA.EQ.1.0.AND.ALPHA.EQ.2.0) GO TO 30
           IF(BETA.EQ.2.0.AND.ALPHA.EQ.1.0) GO TO 40
           GO TO 500
      10   X(1) = 10. ** Y
           GO TO 50
      20   Y = 10.** (2.*Y)
           Z = F * C(2)/(2.* CB)
           X(1) = -Z + (Z** 2. +Y) ** 0.5
           GO TO 50
      30   Y = 10.** (3.* Y)
           Z = F* C(2)/CB
           X(1) = 1.
       8   N= (X(1) ** 3. +Z* X(1) ** 2.-Y)/(3.* X(1) ** 2.+2. *Z * X(1))
           X(1) = X(1)-N
           IF (N-0.0005) 50,8,8
      40   Y = 10.** (3.* Y)
           Z = F * C(2)/CB
           X(1) = 1.
       9   N = (X(1) ** 3. +2.* Z * X(1) ** 2. +X(1) * Z ** 2.-Y)(3.* X(1) **
          1 2. +4. * Z * X(1) + Z ** 2.)
           X(1) = X(1) - N
           IF(N-0.0005) 50,9,9
C          COMPOSITION OF THE SOLUTION SATURATED WITH RESPECT TO (1):
      50   C(1) = X(1) * CB
           A(1) = C(1) * M(1)/10.0
```

```
              P(1) = 100.* A(1)/(100.+A(1)+A(2))
              P(2) = 100.* A(2)/(100.+A(1)+A(2))
              WRITE (6,1004) P(2), A(2), C(2), P(1), A(1), C(1), X(1)
      1004    FORMAT (1H,7F10.4)
              GO TO 1
      500     STOP
              END
```

The resulting data are printed in a table with the heading:

SOLUBILITY OF (1) IN THE TERNARY SYSTEM

M (1) =	M(2) =	α =	β =	F =	Q(1) =	
Q(2) =	AB =	or PB =				
P(2)	A(2)	C(2)	P(1)	A(1)	C(1)	X(1)
...	...	...	...	...	...	...

Using these programs, solubilities were evaluated in a number of ternary systems, the data for which were found in the Linke tables. Data on these systems can be found in Chapter 10.

Example: The KCl solubility is to be calculated in the potassium chloride (1) — ammonium chloride (2) — sodium chloride (3) — water (0) system at 20 °C, if the appropriate data on the ternary systems are known. In the Table 5.5, manual calculation of the constants Q in the potassium chloride (1) — ammonium

Table 5.5. Manual calculation of constants Q in the potassium chloride (1) ammonium chloride (2) water (0) system

a_1	a_2	x_1	x_2	φ_1	m_2	φ_1/m_2
38.10	0.00	1.000	0.000	0.0000	0.00	—
33.64	5.77	0.883	0.139	−0.0073	1.00	−0.007
28.73	13.79	0.754	0.332	−0.0112	2.58	−0.004
24.77	20.02	0.650	0.481	−0.0231	3.75	−0.006
22.99	19.55	0.603	0.470	−0.0496	3.66	−0.014
17.83	28.00	0.468	0.673	−0.0777	5.24	−0.015
16.85	30.22	0.442	0.726	−0.0847	5.65	−0.015
16.57	34.93	0.435	0.840	−0.0636	6.54	−0.010
16.44	28.15	0.432	0.677	−0.0997	5.27	−0.019
15.63	33.88	0.410	0.814	−0.0847	6.34	−0.013
15.25	30.69	0.400	0.738	−0.1073	5.74	−0.019

chloride (2)—water (0) system at 30 °C is shown. The tabulated values, a_1 and a_2 (concentrations in grams per 100 g of water), are given in the first two columns of the table. Using the relationship

$$x_i = \frac{a_i}{a_{0i}} = \frac{p_i}{100 - \sum_i p_i} \cdot \frac{100 - p_{0i}}{p_{0i}}.$$

the values given in the third and fourth columns of the table are calculated from the above values and from them the φ_1 values are found (by using eqn. 5.194), given in the fifth column. The m_2 values were calculated from Table 2.1; the last column contains fractions φ_1/m_2. These values, ordered according to increasing m_2 values, do not exhibit a pronounced dependence and therefore an average value is sufficient, i.e., $Q_{12} = -0.013$. Hence the solubility of potassium chloride in this system is described by the equation

$$\frac{1}{2} \log[x_1(x_1 + 1.525x_2)] = -0.013m_2.$$

This equation expresses the experimental data with the following deviations (the x_1 values were back-calculated on an IBM 360 computer):

| x_2 | x_1 | | Δx_1 |
	Exper.	Calculated	
0.000	1.000	1.000	0.000
0.139	0.883	0.868	−0.015
0.332	0.754	0.707	−0.047
0.481	0.650	0.599	−0.051
0.470	0.603	0.607	0.004
0.673	0.468	0.484	0.016
0.726	0.442	0.456	0.014
0.840	0.435	0.402	−0.033
0.677	0.432	0.482	0.050
0.814	0.410	0.413	0.003
0.738	0.400	0.450	0.050

The solubility of potassium chloride in the potassium chloride(1)—sodium chloride(3)—water(0) system at 25 °C was evaluated in an analogous manner:

$$\frac{1}{2} \log [x_1(x_1 + 1.275x_3)] = -0.014m_3.$$

Therefore, the solubility of potassium chloride in the potassium chloride(1) — ammonium chloride (2) — sodium chloride (3) — water (0) system will be described by the equation

$$\frac{1}{2} \log [x_1(x_1 + 1.525x_2 + 1.275x_3)] = -0.013m_2 - 0.014m_3.$$

If the salt concentrations found experimentally are substituted into the equation, i.e.,

$$m_2 = 3.717 \qquad a_{01} = 34.2$$
$$m_3 = 4.061 \qquad a_{02} = 37.5$$
$$a_{03} = 35.6$$

the relationship

$$x_1^2 + 1.736x_1 - 0.616 = 0$$

is obtained, with the solution $x_1 = 0.3022$. The calculated solubility, $a_1 = 34.2 \times 0.3022 = 10.34$, can be compared with the experimental value, $a_1 = 10.81$.

Example: Solid solutions are formed in the system potassium terephthalate (KT,1)—ammonium terephthalate (NT,2)—water (0). Table 5.6 summarizes the data on the equilibrium phase compositions at 30 °C.

Table 5.6. Phase equilibrium in the system KT - NT - water

No	Solution					Crystals			
	p_1''	p_2''	p_0''	m_1	m_2	p_1'	p_2'	y_1	y_2
1	0.00	10.00	90.00	0.000	0.556	0.00	0.00	0.0000	1.0000
2	1.44	9.20	89.36	0.066	0.515	0.06	99.94	0.0005	0.9995
3	2.26	8.81	88.93	0.105	0.495	0.20	99.80	0.0016	0.9984
4	2.98	8.47	88.55	0.139	0.478	0.44	99.56	0.0036	0.9964
5	3.64	8.24	88.12	0.170	0.468	0.78	99.22	0.0064	0.9936
6	4.25	8.05	87.70	0.200	0.459	1.25	98.75	0.0104	0.9896
7	4.83	7.90	87.27	0.228	0.453	1.88	98.12	0.0156	0.9844
8	5.40	7.74	86.86	0.256	0.446	2.70	97.30	0.0224	0.9776
9	5.92	7.60	86.48	0.282	0.439	3.50	96.50	0.0288	0.9712
10	6.40	7.50	86.10	0.307	0.436	4.40	95.60	0.0361	0.9639
11	6.90	7.40	85.70	0.332	0.432	5.20	94.80	0.0434	0.9566
12	9.08	6.88	84.04	0.446	0.409	11.0	89.00	0.092	0.908
13	11.05	6.40	82.55	0.552	0.388	18.0	82.0	0.155	0.845
14	12.92	5.76	81.32	0.655	0.354	27.0	73.0	0.236	0.764
15	14.64	5.16	80.20	0.753	0.322	39.5	60.5	0.350	0.650
16	16.27	4.36	79.37	0.845	0.275	56.0	44.0	0.510	0.490
17	17.90	3.50	78.60	0.939	0.223	76.5	23.5	0.729	0.271
18	18.96	2.85	78.19	1.000	0.182	86.5	13.5	0.840	0.160
19	19.75	2.40	77.85	1.046	0.154	92.5	7.5	0.912	0.088
20	20.84	1.66	77.50	1.109	0.107	97.5	2.5	1.972	0.028
21	23.10	0.00	76.90	1.239	0.000	100.0	0.0	1.000	0.000

From these data, calculation was first carried out according to eqn. 5.186 (see Table 5.7).

From the last two columns of the table 5.7, it can be seen that φ_1 and φ_2 are dependent on m_2 and m_1, respectively.

Graphical representation of the dependence of φ_1/m_2 on m_2 and of φ_2/m_1 on m_1 showed that the dependence is not simple and that the concentration interval studied should be divided into two parts. The method for calculating the constants is outlined in the Table 5.8 (the graph is not given).

Hence the following equations hold for the solubility of the potassium terephthalate:

$$\frac{1}{3}\log[x_1^2(x_1 + x_2 m_{20}/m_{10})] - \frac{1}{3}\log y_1^2 = -0.75 m_2 + 3.69 m_2^2$$

(valid for $x_2 = 0.00$ to 0.70) and

$$\frac{1}{3}\log[x_1^2(x_1 + x_2 m_{20}/m_{10})] - \frac{1}{3}\log y_1^2 = -5.36 m_2 + 15.0 m_2^2$$

(valid for $x_2 = 0.70$ to 1.00).

Table 5.7. Calculation of constants Q in the system KT-NT-water

No	x_1	x_2	Φ_1	Φ_2	$\frac{1}{3}\log y_1^2$	$\frac{1}{3}\log y_2^2$	φ_1	φ_2
1	0.000	1.000	$-\infty$	-0.0000	$-\infty$	-0.0000	—	-0.0000
2	0.053	0.936	-1.0000	-0.0328	-2.201	-0.0001	1.201	-0.0327
3	0.085	0.890	-0.8410	-0.0454	-1.864	-0.0005	1.023	-0.0449
4	0.112	0.860	-0.7390	-0.0572	-1.631	-0.0011	0.892	-0.0561
5	0.137	0.842	-0.6666	-0.0592	-1.462	-0.0019	0.795	-0.0573
6	0.161	0.826	-0.6178	-0.0614	-1.322	-0.0030	0.704	-0.0584
7	0.184	0.815	-0.5740	-0.0592	-1.205	-0.0045	0.631	-0.0547
8	0.207	0.802	-0.5395	-0.0646	-1.100	-0.0065	0.560	-0.0581
9	0.228	0.790	-0.5075	-0.0603	-1.028	-0.0085	0.520	-0.0518
10	0.248	0.784	-0.4771	-0.0568	-0.962	-0.0107	0.485	-0.0461
11	0.268	0.777	-0.4522	-0.0536	-0.908	-0.0129	0.456	-0.0407
12	0.360	0.736	-0.3487	-0.0526	-0.691	-0.0279	0.342	-0.0247
13	0.446	0.698	-0.2740	-0.0564	-0.540	-0.0487	0.266	-0.0077
14	0.529	0.637	-0.2092	-0.0883	-0.418	-0.0779	0.209	-0.0104
15	0.608	0.579	-0.1646	-0.1256	-0.304	-0.1247	0.139	-0.0009
16	0.682	0.495	-0.1256	-0.2042	-0.195	-0.2131	0.069	$+0.0089$
17	0.758	0.401	-0.0895	-0.3150	-0.091	-0.3780	0.002	0.0630
18	0.807	0.327	-0.0689	-0.4297	-0.051	-0.5301	-0.018	0.1004
19	0.844	0.277	-0.0537	-0.5200	-0.027	-0.7040	-0.027	0.1840
20	0.895	0.192	-0.0350	-0.7280	-0.008	-1.0350	-0.027	0.3070
21	1.000	0.000	0.0000	$-\infty$	0.000	$-\infty$	0.000	—

Table 5.8. Calculation of constants in the system KT-NT-water

No	$\dfrac{\varphi_1/m_2 + 0.75}{m_2}$	$\dfrac{\varphi_1/m_2 + 5.36}{m_2}$	$\dfrac{\varphi_2/m_1 + 0.6}{m_1}$	$\dfrac{\varphi_2/m_1 + 0.1}{m_1^2} + \dfrac{0.15}{m_1}$
1	—	—	—	—
2	—	14.9	1.59	—
3	—	15.0	1.64	—
4	—	15.1	1.42	—
5	—	15.1	1.55	—
6	—	15.0	1.54	—
7	—	14.9	1.58	—
8	—	14.8	1.46	—
9	—	14.9	1.48	—
10	—	14.8	1.47	—
11	—	14.9	1.44	0.40
12	(3.89)	15.2	(1.22)	0.49
13	3.71	15.6	—	0.49
14	3.78	—	—	0.43
15	3.67	—	—	0.44
16	3.64	—	—	0.39
17	3.77	—	—	0.40
18	3.57	—	—	0.47
19	3.69	—	—	0.41
20	(4.67)	—	—	0.44
21	—	—	—	—
average	3.69	15.0	1.48	0.45

The solubility of ammonium terephthalate is described by the equations

$$\frac{1}{3} \log [x_2^2(x_2 + x_1 m_{10}/m_{20})] - \frac{1}{3} \log y_2^2 = -0.60 m_1 + 1.48 m_1^2$$

(valid for $x_1 = 0.0$ to 0.30) and

$$\frac{1}{3} \log [x_2^2(x_2 + x_1 m_{10}/m_{20})] - \frac{1}{3} \log y_2^2 = -0.10 m_1 + 0.15 m_1^2 + 0.45 m_1^3$$

(valid for $x_1 = 0.30$ to 1.00).

These equations contain three variables: the potassium terephthalate concentration in solution, the ammonium terephthalate concentration in solution and the solubility in the solid phase (it simultaneously holds that $y_1 + y_2 = 1$). If the value of one variable is chosen, it is possible, in principle, to calculate the corresponding compositions of the liquid and solid phases by solving the two equations; however, this is difficult and therefore the composition of the liquid phase is expressed independently. In view of the validity of eqns. 5.174 and 5.185 or 5.186, it can be assumed that the empirical expansion of functions $\Phi_1(m_2)$ and $\Phi_2(m_1)$ will be analogous to the expansion of functions φ according to eqn. 5.202. The constants are calculated in Table 5.9. in an analogous manner to the previous case:

Table 5.9. Calculation of constants in the system KT-NT-water

No	$\dfrac{\Phi_1/m_2 + 0{,}23}{m_2}$	$\dfrac{\Phi_1/m_2 - 3.2}{m_2}$	$\dfrac{\Phi_2/m_1 + 0.57}{m_1}$	$\dfrac{\Phi_2/m_1 + 0.50}{m}$	$\dfrac{1.5}{m_1}$
1	—	—	—	—	
2	—	−9.98	1.11	—	
3	—	−9.76	1.31	—	
4	—	−9.93	1.14	—	
5	—	−9.88	1.31	—	
6	—	−9.90	1.32	—	
7	—	−9.86	1.19	—	
8	—	−9.89	1.24	—	
9	—	−9.92	1.26	—	
10	—	−9.85	1.25	—	
11	—	−9.83	1.23	−1.44	
12	—	−9.91	(1.01)	−1.44	
13	(−1.23)	−10.07	—	−1.41	
14	−1.02	—	—	−1.44	
15	−0.87	—	—	−1.41	
16	−0.83	—	—	−1.41	
17	−0.77	—	—	−1.41	
18	−0.82	—	—	−1.43	
19	−0.77	—	—	−1.43	
20	−0.91	—	—	−1.48	
21	—	—	—	—	
average	−0.86	−9.90	1.24	−1.43	

Hence the resulting equations have the following form:
Solubility of potassium terephthalate:

$$\frac{1}{3}\log x_1^2(x_1 + x_2 m_{20}/m_{10})] = -0.23 m_2 - 0.86 m_2^2$$

(valid for $x_2 = 0.00$ to 0.65) and

$$\frac{1}{3}\log [x_1^2(x_1 + x_2 m_{20}/m_{10})] = 3.20 m_2 - 9.9 m_2^2$$

(valid for $x_2 = 0.65$ to 1.00).

The solubility of ammonium terephthalate:

$$\frac{1}{3}\log [x_2^2(x_2 + x_1 m_{10}/m_{20})] = -0.57 m_1 + 1.24 m_1^2$$

(valid for $x_1 = 0.00$ to 0.30) and

$$\frac{1}{3}\log [x_2^2(x_2 + x_1 m_{10}/m_{20})] = -0.50 m_1 + 1.50 m_1^2 - 1.43 m_1^3$$

(valid for $x_1 = 0.30$ to 1.00).

By subtracting the appropriate equations, $\Phi(m)$, from the above given series, $\varphi(m)$, a third series of equations is obtained, giving the equilibrium composition of the solid phase:

$$\log y_1 = 0.78\, m_2 - 6.83\, m_2^2 \qquad (*)$$

(valid for $m_2 < 0.4$) and

$$\log y_2 = 0.045\, m_1 - 0.36\, m_1^2$$

(valid for $m_1 < 0.3$). $\qquad (**)$

The measured and calculated data are compared in the Table 5.10.

Another modification of the method described above is based on eqn. 5.193 or 5.194, taking antilogarithms

$$x_1^{\nu_{1-}/\nu_1}\left(\sum_{i=1}^{n}\frac{\nu_{i+} + m_{i0}}{\nu_{1+} + m_{10}} x_i\right)^{\nu_{1+}/\nu_1} = \psi_1 \qquad (5.207)$$

for a common cation, or

$$x_1^{\nu_{1+}/\nu_1}\left(\sum_{i=1}^{n}\frac{\nu_{i-} - m_{i0}}{\nu_{1-} - m_{10}} x_i\right)^{\nu_{1-}/\nu_1} = \psi_1 \qquad (5.208)$$

for a common anion. For the function

$$\psi_1 = 10^{\varphi_1} \qquad (5.209)$$

an expansion formally identical with eqn. 5.202 is used, i.e.,

$$\psi_1 = 1 + \sum_{i=2}^{n} A_{1i} x_i + \sum_{i=2}^{n}\sum_{j=2}^{n} A_{1ij} x_i x_j + \ldots \qquad (5.210)$$

Table 5.10. Calculation of phase equilibria in the system KT-NT-water

No	m_2	m_1	y_1 Eq. (∗)	y_1 Eq. (∗∗)	y_1 exper.	x_1
1		0.000		0.000	0.000	0.000
2		0.066		0.000	0.001	0.053
3		0.105		0.000	0.002	0.085
4		0.139		0.002	0.004	0.112
5		0.170		0.006	0.006	0.137
6		0.200		0.013	0.010	0.161
7		0.228		0.019	0.016	0.184
8		0.256		0.027	0.022	0.207
9		0.282		0.036	0.029	0.228
10		0.307		0.045	0.036	0.248
11	0.432	0.332	0.115	0.054	0.043	0.268
12	0.409	0.446	0.150	0.112	0.092	0.360
13	0.388		0.188		0.155	0.446
14	0.354		0.263		0.236	0.529
15	0.322		0.349		0.350	0.608
16	0.275		0.499		0.510	0.682
17	0.223		0.683		0.729	0.758
18	0.182		0.824		0.840	0.807
19	0.154		0.908		0.912	0.844
20	0.107		1.000		0.972	0.895
21	0.000		1.000		1.000	1.000

with the difference that the expansion is carried out with respect to relative molalities. It holds again for interaction constants A that their temperature dependence is not large and that they can be combined for equations describing multi-component systems. In order to assess higher constants, the relationships

$$A_{1ij} = A_{1ji} \tag{5.211}$$

$$A_{1ij} = (A_{1ii} \cdot A_{1jj})^{1/2} \tag{5.212}$$

can be used. If eqns. 5.207 and 5.208 are rewritten in three basic forms, viz.,

$$\psi_1 = x_1(x_1 + B) \tag{5.213}$$

$$\psi_1 = x_1^2(x_1 + B) \tag{5.214}$$

and

$$\psi_1 = x_1(x_1 + B)^2 \tag{5.215}$$

where

$$B = \sum_{i=2}^{n} \frac{v_i \mp m_{i0}}{v_1 \mp m_{10}} x_i \tag{5.216}$$

the solution of these equations can then be simplified by using Table 5.11.

Example: The solubility of sodium chloride in the sodium chloride (1)—potassium chloride (2) — water (0) system in the temperature range 0—100 °C can be expressed by the equation

$$x_1 \left(x_1 + \frac{m_{20}}{m_{10}} x_2 \right) = 1 - 0.012 \, x_2 \, .$$

Table 5.11. The solution of eqns. 5.213—5.2.15; φ_1 is given as a function of x_1 and B

Equation	x_1	$B=0$	0.5	1.0	1.5	2.0	2.5	3.0	3.5	4.0	4.5	5.0
5.213	0.1	0.01	0.06	0.11	0.16	0.21	0.26	0.31	0.36	0.41	0.46	0.51
	0.2	0.04	0.14	0.24	0.34	0.44	0.54	0.64	0.74	0.84	0.94	1.04
	0.3	0.09	0.24	0.39	0.54	0.69	0.84	0.99	1.14	1.29	1.44	1.59
	0.4	0.16	0.36	0.56	0.76	0.96	1.16	1.36	1.56	1.76	1.96	2.16
	0.5	0.25	0.50	0.75	1.00	1.25	1.50	1.75	2.00	2.25	2.50	2.75
	0.6	0.36	0.66	0.96	1.26	1.56	1.86	2.16	2.46	2.76	3.06	3.36
	0.7	0.49	0.84	1.19	1.54	1.89	2.24	2.59	2.94	3.29	3.64	3.99
	0.8	0.64	1.04	1.44	1.84	2.24	2.64	3.04	3.44	3.84	4.24	4.64
	0.9	0.81	1.26	1.71	2.16	2.61	3 06	3.51	3.96	4.41	4.86	5.31
	1.0	1.00	1.50	2.00	2.50	3.00	3.50	4.00	4.50	5.00	5.50	6.00
5.214	0.1	0.00	0.01	0.01	0.02	0.02	0.03	0.03	0.04	0.04	0.05	0.05
	0.2	0.01	0.03	0.05	0.07	0.09	0.11	0.13	0.15	0.17	0.19	0.21
	0.3	0.03	0.07	0.12	0.16	0.21	0.25	0.30	0.34	0.39	0.43	0.48
	0.4	0.06	0.14	0.22	0.30	0.38	0.46	0.54	0.62	0.70	0.78	0.86
	0.5	0.13	0.25	0.38	0.50	0.68	0.75	0.88	1.00	1.13	1.25	1.38
	0.6	0.22	0.40	0.58	0.76	0.94	1.12	1.30	1.48	1.66	1.84	2.02
	0.7	0.34	0.59	0.83	1.08	1.32	1.57	1.81	2.06	2.30	2.55	2.79
	0.8	0.51	0.83	1.15	1.47	1.79	2.11	2.43	2.75	3.07	3.39	3.71
	0.9	0.73	1.13	1.54	1.94	2.35	2.75	3.16	3.56	3.97	4.37	4.78
	1.0	1.00	1.50	2.00	2.50	3.00	3.50	4.00	4.50	5.00	5.50	6.00
5.215	0.1	0.00	0.04	0.12	0.26	0.44	0.68	0.96	1.30	1.68	2.12	2.60
	0.2	0.01	0.10	0.29	0.58	0.97	1.46	2.05	2.74	3.53	4.42	5.41
	0.3	0.03	0.24	0.51	0.97	1.59	2.35	3.27	4.33	5.55	6.91	8.43
	0.4	0.06	0.32	0.78	1.44	2.31	3.36	4.62	6.08	7.74	9.60	11.66
	0.5	0.13	0.50	1.13	2.00	3.13	4.50	6.13	8.00	10.13	12.50	15.13
	0.6	0.22	0.73	1.54	2.65	4.06	5.77	7.78	10.09	12.70	15.61	18.82
	0.7	0.34	1.01	2.02	3.39	5.10	7.17	9.58	12.35	15.46	18.93	22.74
	0.8	0.51	1.35	2.59	4.23	6.27	8.71	11.55	14.79	18.43	22.47	26.91
	0.9	0.73	1.76	3.25	5.18	7.57	10.40	13.69	17.42	21.61	26.24	31.33
	1.0	1.00	2.25	4.00	6.25	9.00	12.25	16.00	20.25	25.00	30.25	36.00

The solubility of sodium chloride in the sodium chloride (1) — magnesium chloride (3)—water (0) system at 25 °C can be expressed in an analogous manner by the equation

$$x_1 \left(x_1 + 2 \frac{m_{30}}{m_{10}} x_3 \right) = 1 - 1.938\, x_3 + 0.962\, x_3^2 .$$

By using these two equations, the solubility of sodium chloride in the four-component system sodim chloride (1) — potassium chloride (2) — magnesium chloride (3) — water (0) can be calculated:

$$x_1 \left(x_1 + \frac{m_{20}}{m_{10}} x_2 + \frac{m_{30}}{m_{10}} x_3 \right) = 1 - 0.012 x_2 - 1.938\, x_3 + 0.962\, x_3^2 .$$

It has been found that these equations express the experimental data with a mean quadratic deviation of about 0.4 %.

The correlation method described in this chapter has the following properties:

(a) It enables the data to be presented conveniently in a compact algebraic equation for an arbitrary number of components.

(b) The equations are not bound by any special electrolyte solution theory (although some connecting points can be found) and therefore they are free from the disadvantages of a theoretical treatment.

(c) The precision of the data can be improved to equal that of the experimental determination. The number of variable parameters necessary for expressing the solubility in a ternary system is usually 1 or 2, which enables the amount of experimental data to be reduced to a minimum.

(d) The coefficients of the resulting equations exhibit only a small temperature dependence, so that they can be used for temperature interpolations and, to a limited extent, for extrapolations.

(e) It is possible that the interaction constants, Q, could be correlated with the properties of the components of the system after a sufficient number of systems have been handled, i.e. to generalize.

(f) The method permits the calculation of solubilities in multicomponent systems, based on the behaviour of partial three-component systems.

(g) It has been found that the classical mass action law can be derived not only under the very specialized assumption of an ideal solution, but also under the much broader assumption that the activity coefficient of a pure electrolyte in solution equals the activity coefficient of the same electrolyte in a mixture.

6. IMPURITY INCLUSION IN CRYSTALS

From the previous chapters, especially from Chapter 3, the reader might obtain the impression that, in systems that do not form solid solutions, the component crystals are always perfectly pure. Unfortunately, this need not necessarily be so and there are many reasons for the contamination of crystals. If contamination due to mother liquor adhering to the crystal surface, which can be removed by careful washing of the product, is ignored, there is still a theoretical possibility that has not been discussed here: the extent to which it is possible to distinguish type 2-I systems whose components are completely immiscible in the solid phase from type 2-III systems with components that are partially (and possibly only slightly) miscible in the solid phase.

In this chapter, the extensive literature dealing with individual systems will not be treated, and only the basic mechanisms of impurity inclusion into crystals[1] will be briefly described.

6.1. Isomorphous inclusion

According to the Hahn precipitation rule, coprecipitation of micro-impurities in crystals of the macro-component always occurs when the micro-component is included isomorphously into the crystal lattice of the macro-component or when contributes to normal lattice formation. Processes in which this rule plays role are characterized by the independence of the coprecipitation process of the crystallisation rate and of the amount of substance crystalised and by the formation mixed crystals. Therefore, systems in which isomorphous impurity inclusion urs are usually of the 2-II or 2-III type.

vo basic distribution functions have been formulated for isomorphous impurity sion, as follows.

rner and Hoskins assume that, during crystal growth, an exchange reaction n macro-component particles A, contained in the lattice, and micro- ent particles in solution occurs on the surface of each newly formed layer. ig that the crystallisation is very slow, equilibrium is established in the ind can be described by the Guldberg-Waage law in the form

$$\frac{[X]_s}{[A]_s} = \frac{K[X]_l}{[A]_l} \qquad (6.1)$$

bilský, J. Nývlt and L. Jäger, Chem. Průmysl, 18 (1968) 14.

where [X] and [A] denote the concentrations of the micro- and macro-component, respectively, and subscripts s and l relate to the surface of the solid phase and the solution volume phase, respectively. If the initial micro-component concentration is x_0 and that of the macro-component is y_0 and if their amounts in the solution after a certain crystallisation time are $(x_0 - x)$ and $(y_0 - y)$, where x is the separated part of the micro-component and y that of the macro-component, then, assuming that y is sufficiently small,

$$\frac{dx}{dy} = \lambda \frac{x_0 - x}{y_0 - y} \qquad (6.2)$$

where λ is a proportionality constant. On intergration of eqn. 6.2 for y from 0 to y and for x from 0 to x, the relationship

$$\log \frac{x_0}{x_0 - x} = \lambda \log \frac{y_0}{y_0 - y} \qquad (6.3)$$

is obtained, which is known as the logarithmic distribution law, where the constant λ is the logarithmic distribution coefficient.

If $\lambda > 1$, the solid phase is enriched in the micro-component during the crystallisation; otherwise, the solid phase is depleted in the micro-component. In addition to equilibrium between each crystal layer formed and the solution, the logarithmic law assumes the impossibility of mass exchange between the interior of the crystal. and its surface, i.e., it assumes that diffusion in the solid phase is negligible. Therefore, the component distribution in the solid phase is inhomogeneous: for $\lambda > 1$, the highest micro-component concentration is in the centre of the crystal and it decreases towards the surface, while for $\lambda < 1$ this distribution is reversed.

The Nernst distribution assumes that, during crystallization, the ions of the two components are in equilibrium with the ions inside the crystal. The micro-component distribution is then homogeneous throughout the crystal and the thermodynamic condition of equality of the chemical potentials of the micro-component in the two phases can be employed. It follows from eqns. 5.161—5.164 that

$$\ln\left[\frac{a_x''}{a_x'}\right] = \frac{\mu_x^{0'} - \mu_x^{0''}}{RT} \qquad (6.4)$$

where a_x'' and a_x' are the activities of the micro-component in the liquid and solid phases, respectively. As the standard chemical potentials are constant at a given temperature, then

$$\frac{a_x''}{a_x'} = \text{const.} \qquad (6.5)$$

and this thermodynamically rigorous relationship can be approximated by the equations

$$\frac{x}{y} = D \cdot \frac{x_0 - x}{y_0 - y}. \qquad (6.6)$$

and
$$\frac{x}{x_0 - x} = D \cdot \frac{y}{y_0 - y}. \tag{6.7}$$

In these equations, y_0 and x_0 denote the initial concentrations of the two components, x and y are their corresponding amounts in the solid phase and D is the homogeneous distribution coefficient.

The homogeneous distribution law states that, when two substances separate in the form of isomorphous mixed crystals, the micro- and macro-components are distributed between the solid and liquid phases in a constant ratio. For $D > 1$, the solid phase is enriched in the micro-component, while for $D < 1$ it is depleted, compared with the micro-component contained in the solution.

A general solution of the distribution law for isomorphous replacement can also be found in the literature. It follows from the general solution that:

(a) Homogeneous distribution holds for a constant concentration of the macro- and micro-components in solution (eqn. 6.7).

(b) For a constant concentration of macro-component and a variable concentration of micro-component, eqn. 6.3 is obtained for non-uniform distribution.

(c) If neither the concentration of macro-component nor that of the micro-component is constant, complicated expressions are obtained that can be solved for slow equilibrium crystallisation. Employing a number of simplifying assumptions, equations of the eqn. 6.3 or 6.7 type are obtained.

6.2. Adsorption inclusion

Syncrystallisation of an impurity is often dependent on the structural properties of the crystals of the macro-component and its extent can be controlled by the external conditions of the solid phase separation.

Adsorption inclusion can be characterized very roughly by the Paneth rule: the cation of an element is more strongly adsorbed on the separating solid phase, the less soluble is the component that it forms with the anion of the solid phase. The polarization effect of adsorbed ions also plays a role here, so that adsorption can take place even on a neutrally or identically charged surface. In any event, a dependence between the amount of component adsorbed (micro-component) and the specific surface of the macro-component crystals can be found.

6.3. Formation of anomalous mixed crystals

Sometimes (e.g., in syncrystallisation of heavy metal chlorides with ammonium chloride), true mixed crystals cannot be formed as the micro- and macro-component are not isomorphous owing to their chemical nature and crystal structure parameters. Adsorption is also not involved, as the micro-component is distributed homogeneously throughout the crystals and the distribution coefficient is independent of the external crystallisation conditions and of the specific surface of the crystal. With this type of system, a so called lower miscibility limit can often be observed; on decreasing the concentration of the micro-component in the solution below a certain limit, syncrystallisation of the two components suddenly stops. This phenomenon can be explained by the assumption that, while in isomorphous inclusion the particles of the macro-component are replaced with particles of the micro-component in the lattice being formed, here whole sections of the lattice are replaced. At very low concentrations of micro-component, the formation of such lattice sections on the crystal surface is improbable and hence a lower miscibility limit appears.

6.4. The mechanism of internal adsorption

Internal adsorption is intermediate between isomorphous inclusion and adsorption. In a similar manner to adsorption, it is characterized by variability of the distribution coefficient with changes in the crystallisation conditions; on the other hand, the micro-component is also contained inside the crystals of the solid phase. The crystals exhibit a regular discontinuous distribution of the micro-component, owing to selective adsorption on certain faces of the growing crystal: when the distribution of impurity in the crystal is monitored, the crystal is found to be subdivided into individual sectors, identical with bipyramids of faces that exhibit selective adsorption. This phenomenon is therefore also called sectorial crystal growth. The existence of such selective adsorption is also reflected in the shapes of crystals obtained from a solution in the presence and absence of the micro-component, these shapes usually being different.

6.5. Mechanical inclusions

Under certain conditions, trace amounts of substances in solutions can form colloids, the centres of which can be formed by various impurities (dust, etc.). During separation of the solid phase, these substances can be deposited on crystal faces and covered by subsequent layers of the growing crystal. Another extreme case is inclusion of the mother liquor inside the growing crystals which usually occurs when crystallisation proceeds relatively rapidly in unstirred solutions.

7. LITERATURE

It is not the aim of this monograph to present an exhaustive survey of the literature in the given field. Therefore, only basic work that was employed as a direct source for the individual chapters is shown. Other references can be found in the cited literature.

E. Hála and A. Reiser, *Physical Chemistry* (in Czech), NČSAV, Prague, 1960.
A group of workers in the physical chemistry department of the Institute of Chemical Technology in Prague: *Physical Chemistry I* (a textbook in Czech), SNTL, Prague 1962 (Chapter 4 by V. Fried and J. Pick).
S. Glasstone, *Textbook of Physical Chemistry*, Van Nostrand, New York, 1947.
J. Nývlt, *Crystallisation from Solutions* (in Slovak), SVTL, Bratislava, 1967.
J. Nývlt: *Industrial Crystallisation from Solutions*, Butterworths, London, 1971.
J. W. Mullin, *Crystallisation*, Butterworths, London, 1972.
O. D. Kashkarov, *Graficheskie Raschety Solevykh Sistem* (in Russian — Graphical Evaluation of Salt Systems), GKhI, Leningrad, 1960.
H. Lux, *Anorganisch — chemische Experimentierkunst*, Barth, Leipzig, 1959.

8. TABLES OF SOLUBILITIES OF VARIOUS SUBSTANCES IN WATER

Binary systems

The solubility (wt. — % of various substances in the anhydrous form) and the first heats of solution (kJ/mole) for compounds in water. The temperatures of hydrate changes are given in parentheses.

Substance	H₂O	\multicolumn{11}{c}{t °C}	ΔH_∞
		0	10	20	30	40	50	60	70	80	90	100	
AgCH₃COO	0	0.71	0.87	1.03	1.20	1.39	1.61	1.85	2.13	2.46			−18.4
AgBrO₃	0				0.227	0.316	0.433	0.570	0.735	0.936	1.325		
AgClO₃	0		8.5	13.0	17.0								−31.4
AgClO₄	1 0	81.3	82.9	84.0	85.6	(43°)	87.2	87.5	87.8	88.3	88.5	88.85	+9.2
AgF	4 2 0	46.18 27.50	54.5	(18.65°) 63.23	65.53	(39.5°)	68.35						−20.5 +18.0
AgNO₂	0	0.15	0.22	0.34	0.50	0.71	0.99	1.36					−36.8
AgNO₃	0	54.8	62.6	68.4	72.6	75.7	78.9	81.5	82.8	85.4	86.7	88.0	−22.6
Ag₂SO₄	0	0.57	0.69	0.79	0.88	0.97	1.05	1.14	1.21	1.28	1.34	1.39	−18.8
AlCl₃	6	31.0	31.66	31.81	31.96	31.63		31.73	32.46	32.74		33.25	+54.8
Al₂(SO₄)₃	18 16	23.8 27.50	25.1 27.6	26.7	28.8 28.0	31.4 28.8	34.3 29.9	37.2 31.0	32.8	42.2	39.6	47.1 44.7	+34.8
AlF₃	3	0.25	0.28										−7.1
AlK(SO₄)₂	12	3.0	4.0	5.9	8.39	11.70	17.0	24.75	40.0	71.0	109.0		−42.3
Al(NH₄)₃F₆	0	0.43					1.23	1.17	1.11			0.74	
Al(NH₄)(SO₄)₂	12	2.97	4.13	5.91	8.26	12.0	18.2	27.7	44.3	86.2			−38.5
Al(NO₃)₃	9 8	37.5	40.0	42.5	45.0	47.0 —	49.0	51.5	54.0	57.0	(90°) 60.5	61.5	
AlNa₃F₆	0	0.034					0.040					0.135	

AlNa(SO$_4$)$_2$	12	27.24	28.23	28.43	29.45					
As$_2$O$_3$	0	1.2	1.4	1.8	2.28	2.8	3.45	4.3	5.0	5.7	6.6	7.6	+25.1
As$_2$O$_5$	4 5/3	37.3	38.3	39.7	(29.5°) 41.6			42.2	42.55	42.9		43.3	−17.2
AuKCl$_4$	0		27.7	38.2	48.7	59.2	70.0	80.2					−63.2
AuNaCl$_4$	2		58.2	60.2	64.0	69.4	77.5	90.0					−16.7
Ba(OOCCH$_3$)$_2$	3 2 0	37.0	38.7	41.6	(24.7°) 42.7	43.9	(41°) 43.5	42.9	42.6	42.45	42.6	42.8	−46.9
BaBr$_2$	2	47.5	48.5	49.5	50.6	51.3	52.5	53.5	54.5	55.5	56.6	57.8	−10.0
Ba(BrO$_3$)$_2$	1	0.28	0.43	0.65	0.95	1.31	1.72	2.27	2.92	3.52	4.26		
BaCl$_2$	2	23.8	25.0	26.3	27.7	28.9	30.3	31.6	33.1	34.3	35.7	37.0	−36.0
Ba(ClO$_3$)$_2$	1	16.90	21.23	25.26	29.43	33.16		40.05		45.90		51.2	−42.7
Ba(OOCH)$_2$	0	20.8	21.9	23.0	24.2	25.4	27.9	29.3	30.8	31.3	32.3	33.9	−60.7
BaI$_2$	15/2 2 1	62.5	64.8	67.15	(25.7°) 64.1	69.6	70.1	70.7	71.25	71.96	72.7	(98.9°)	+30.1
Ba(NO$_2$)$_2$	1			40.3	44.3			58.5		67.3			
Ba(NO$_3$)$_2$	1	4.72	6.25	8.27	10.3	12.35	14.6	16.9	19.1	21.4	23.6	25.6	
Ba(OH)$_2$	8	1.65	2.42	3.74	5.29	7.60	11.61	17.32		50.35			
BaS	0	2.80	4.66	7.28	9.40	12.96	17.62	21.69	27.19	33.29	40.24	37.61	
Ba(SH)$_2$	0	32.6		32.8		34.5		36.2		39.0		43.7	
BaS$_2$O$_3$	0		0.194	0.243	0.288	0.329	0.366	0.399					

contd.

Substance	H₂O	\multicolumn{11}{c}{$t°C$}	ΔH_∞
		0	10	20	30	40	50	60	70	80	90	100	
BaS₂O₆	4 2	6.1	11.1	(13°) 15.7	19.9								
BeCl₂	4	40.35		42.24	43.52	44.12							
Be(NO₃)₂	4	49.6		51.9	52.3		58.6	64.24					
BeSO₄	4	26.1	27.0	28.0	29.1	30.7	32.6	35.1	37.5				+4.6
Ca(OOCCH₃)₂	2 1	21.4	21.0	20.5	20.2	19.9		19.7		20.1	(85°) 19.2	18.6	+24.7
Ca(OOCC₆H₅)₂	3 1	2.25	2.46	2.72	3.02	3.6	4.05	4.71	5.68	6.87	(84.7°) 7.9	8.7	
CaBr₂	6 4	55.5	57.0	58.8	(34.2°) 68.1			73.5		74.7		75.4	+4.6 −30.6
CaCl₂	6 4 2	37.3	39.3	42.7	(30.1°) 50.1	53.4	(41°) 57.0	57.8	58.6	59.5	60.6	61.4	−19.3 +8.0 +41.9
Ca(ClO₃)₂	2 0			66.16	67.0	68.9	70.8	72.7	74.6	(76°) 77.3	77.9	78.5	
CaCrO₄	0	4.3		2.2			1.11	0.82	0.79			0.42	
Ca(OOCH)₂	0	13.90	14.22			14.56		14.89		15.22		15.53	
CaI₂	6	64.6	66.0	67.7	69.0	70.8		74.0		78.0		81.0	
Ca(IO₃)₂	6 1 0	0.090		(30°) 0.384 0.387	(30°) 0.517	0.590	(57.5°) 0.617	0.644	0.665	0.668			

Compound	n												Δ
Ca(NO2)2	4	39.01		46.8	51.7	(34.6°) 55.3	55.9	56.5	58.4	60.3	62.5	64.2	
	1		42.9										
Ca(NO3)2	4	50.5		56.39	60.41	66.22	(42.5°) (49°)						−0.4
	3		53.55			73.8	77.8						−18.0
	2												−14.2
	0												+17.1
Ca(OH)2	0	0.130	0.125	0.118	0.109	0.100	0.091	0.081	(51°)	0.065	0.059	0.052	+12.6
CaSO4	2	0.176	0.193	0.202	0.209	0.210			0.188			0.162	−3.5
	1/2			0.298	0.255	0.22	0.15					0.067	
CaS2O3	6	25.75	29.40	33.05	36.75	40.45							
CdBr2	4	36.0	43.0	49.7	56.3	(36.0°) 60.3	60.35	60.45	60.7			61.65	−30.6
	0												+1.8
CdCl2	2 1/2	47.3			56.91	(33.8°)		57.7		58.35		59.55	−12.6
	1												+2.6
CdI2	0	44.05		45.85	46.79	47.95	50.1			52.6		55.55	−4.0
Cd(NO3)2	9	52.0	(3.5°) 57.7	60.0	62.8	67.0	(56.80°) 86.05		86.5	87.0	87.1	87.2	−21.4
	4							(48.7°)					
	2							82.5					
	0												
CdSO4	8/3	43.0	43.2	43.3	43.5	44.0	42.5	(43.6°)	41.4	40.3		36.9	+10.5
	1							43.5					+25.1
Ce(NO3)3	6				65.72	69.7							−20.9
								73.88					
Ce2(SO4)3	12	14.20	14.6	9.0	6.8	5.6	3.73	4.46	2.16	1.19	0.82	0.45	
	9	17.35	13.0	8.60	7.1	5.61	3.88	4.7	1.60	0.90	0.65	0.41	
	8	15.95	12.2				3.14	6.41					
	5					5.71	2.35	3.31					
	4												

185

contd.

Substance	H₂O	\multicolumn{11}{c}{t °C}	ΔH_∞
		0	10	20	30	40	50	60	70	80	90	100	
CoBr₂	6	47.9	50.5	53.1	57.0	60.9	(43°) 66.8	(60°) 69.4	70.1				+5.4
	4												
	2											72.0	
CoCl₂	6	30.3	32.3	34.6	37.4	41.0	(49°) 46.2	(58°) 48.4	48.8	49.4	50.3	51.5	−12.1
	4												
	2												+41.4
CoI₂	1	58.0	61.5	65.2	70.0	75.0	79.0	79.3	79.6	80.0	80.3	80.6	
Co(NH₄)₂(SO₄)₂	6	9.0	11.12	13.50	16.0	18.94	21.88	25.15	28.44	32.44	37.12	43.56	
Co(NO₂)₂	0	0.076	0.24	0.40	0.60	0.84							
Co(NO₃)₂	6	45.66		49.33	52.7	55.8	60.12	(55°) 62.88	64.89	67.86	77.1		−20.5
	3												
CoSO₄	7	19.8	23.7	25.6	28.7	31.5	(43.3°) 33.8	35.7	(61.30°) 35.4	33.0	30.3	27.8	−15.1
	6												
	1												
CrNH₄(SO₄)₂	12	3.8			15.7	24.5							
Cr(NO₃)₃	9		40.81	43.03	45.5								
CrO₃	0	62.24		62.58		63.3	64.5	64.78	65.1	65.7	66.5	67.4	+10.5
CsAl(SO₄)₂	12	0.21	0.30	0.40	0.60	0.85	1.30	2.0	3.10	5.16	9.5	18.61	
CsCl	0	61.7	63.6	65.1	66.4	67.5	68.6	69.7	70.6	71.4	72.2	73.0	−19.3
CsI	0	30.6	37.3	44.1	49.0	53.2	57.03	60.6	63.1	65.6	67.5	69.6	−35.6
CsNO₃	0	8.54	12.97	18.7	25.3	32.1	39.2	45.6	51.7	57.3	62.0	66.3	−41.4

CsOH	0						75.18	80.8				+69.5
Cs₂SO₄	0	62.6	63.4	64.1	64.8	65.5	66.1	66.7	67.2	67.8	68.3	68.8	−20.5
CuBr₂	4/0	51.8	53.8	(18°) 55.9	56.0	56.4	56.8						−4.2 / +34.8
CuCl₂	2	40.8	41.5	42.2	43.9	44.75	45.0	46.5		49.5	50.9	54.60	+15.5
CuSO₄·(NH₄)₂SO₄	6	10.4	13.10	16.22	19.64	23.34	27.34	31.63	36.21	41.08	46.25	51.71	
Cu(NO₃)₂	6 / 2½	45.5	50.0	55.5	(25.4°) 61.0	62.0	63.2	64.5	66.0	67.5	69.0	71.2	−44.8 / −10.5
CuSO₄	5 / 3	12.3	14.5	16.8	19.4	22.3	25.3	28.5	32.0	35.9	40.3	(95.9°) 43.5	−12.1 / +15.1
FeBr₂		50.54	51.9	53.6	55.4	56.9	(49°) 58.5	59.0	60.5	62.6	(83°) 63.8	64.8	
FeCl₂	6 / 4 / 2	33.2	37.2	(12.3°) 38.5	39.7	40.7	42.2	43.9	45.8	(76.5°) 47.55	48.1	48.7	+11.3 / +36.4
FeCl₃	6 / 3.5 / 2 / 0	42.66		47.88	51.64		75.91 / 78.23	78.86	80.00 / 83.8	84.0		84.2	+23.4 / +132.7
FeSO₄·K₂SO₄	6	16.4	19.6	25.1	27.9	31.1	34.4	36.4	39.09				
Fe(NH₄)₂(SO₄)₂	6	11.1	16.0	20.9	24.58	27.8	28.6	31.9	34.2				−41.0
Fe(NO₃)₂	6	41.53	43.2	45.6	48.0	51.5	55.5	62.4					
Fe(NO₃)₃	9	40.15	43.8	45.2	48.2	51.18							−38.1
FeSO₄	7 / 4 / 1	13.6	17.2	20.8	24.7	28.6	32.6	(56.6°) 35.5	(63.7°) 33.5	30.4	27.2	24.0	−18.8 / +6.7 / +31.4

contd.

Substance	H_2O	t °C										ΔH_∞
		0	10	20	30	40	50	60	70	80	90	100	
H_3BO_3	0	2.70	3.52	4.65	6.34	8.17	10.23	12.96	15.75	19.06	23.27	27.53	−21.4
$HgBr_2$	0	0.3	0.4	0.55	0.65	0.9	1.25	1.65		2.7		4.7	−10.0
$Hg(CN)_2$	0	6.31	8.09	10.45									−12.6
$HgCl_2$*	0	3.5	4.6	6.1	7.7	9.3	11.4	14.4	18.8	23.4	29.6	36.5	−13.8
$InCl_3$	4 3 2.5		63.9	66.1	69.7	(30.5°) 71.52	73.42	75.34	77.71	(80°) 78.87	79.50	80.8	
KBF_4	0			0.44	0.75							5.90	
$KO \cdot OCCH_3$	1.5 0.5	68.4	70.1	71.9	73.9	76.4	(41.3°) 77.1	77.8	78.5	79.2	79.9		
$KO \cdot OCC_6H_5$	0	38.7	40.2	41.7	43.3	44.8	46.6	47.9	49.5	51.0	52.5	54.0	−21.4
KBr	0	34.9	37.3	39.4	41.4	43.2	44.7	46.1	47.4	48.7	49.8	51.0	−42.3
$KBrO_3$	0	2.96	4.5	6.46	8.78	11.58	14.69	18.21		25.53		33.3	−1.9
K_2CO_3	1.5	51.26	51.9	52.5	53.2	53.9	54.8	55.9	57.1	58.3	59.6	60.9	−31.4
$K_2C_2O_4$	1	20.28	23.20	25.95	28.70	31.2	33.5	35.6	37.8	40.2	42.0	44.5	
$K_3C_6H_5O_7$	1			63.2	66.0								
KCl	0	21.92	23.8	25.5	27.1	28.6	30.0	31.4	32.7	33.9	35.0	36.0	−18.4
$KClO_3$	0	3.2	4.9	6.8	9.2	12.2	15.0	19.2	23.2	27.3	31.5	36.0	−43.1
$KClO_4$	0	0.75	1.05	1.65	2.50	3.60	4.90	6.8	9.2	11.8	15.0	18.2	−52.8
K_2CrO_4	0	36.4	37.9	38.9	39.5	40.1	40.8	42.1		44.5		46.5	−20.1

$K_2Cr_2O_7$	0	4.4	6.2	10.5	15.4	20.7	27.2	31.7	36.8	40.8	45.1	48.25	−74.5
KF	4	30.9	34.87	(17.7°) 48.7	51.95	58.0	(40.2°) 58.6						−26.0
	2									60.01			−8.4
	0							58.72	59.4				+16.3
$K_3Fe(CN)_6$	0	23.2	27.4	31.0	34.4	37.6	39.5	41.6	43.4	45.2	46.4	47.8	−60.3
$K_4Fe(CN)_6$	3	12.5	17.36	22.0	26.0	29.3	32.6	34.2	38.2	40.1	(87.3°) 41.6	42.7	−51.9
	0												
KH_2AsO_4	1	15.7	18.3	22.0	25.2	28.0	32.0	(59.6°) 35.0	38.3	41.6	45.0	48.0	+20.1
	0												
K_2HAsO_4	6	48.5	55.5	(17°) 62.7	64.5	67.0	69.7	72.5	(66.5°) 75.1	77.5	(84°) 78.6	79.8	
	3												
	1												
	0												
$KHCO_3$	0	18.6	21.8	25.0	28.1	31.3	34.2	37.5	40.6				−21.4
KHF_2	0	19.7	23.14	28.15				44.08		53.28			−24.3
KH_2PO_4	0	12.88	15.50	18.45	21.90	25.10	29.00	33.40	37.05	41.30	45.50		+19.7
K_2HPO_4	6	45.58	53.8	(14.7°) 60.6	63.4	67.2	(48.3°) 71.95	72.3	72.5	72.7		73.8	
	3												
	0												
$KC_8H_5O_4$	0	60.4			76.65		78.9						
$KHSO_4$	0	26.6	32.7	35.4	37.9	40.3	43.3	46.1	54.9				−13.0
KI	0	56.0	57.6	59.0	60.4	61.6	62.8	63.8	64.8	65.7	66.6	67.4	−21.8
KIO_3	0	4.4	5.9	7.48	9.3	11.2	13.3	15.5	19.9	24.4		−28.9
KIO_4	0	0.16	0.25	0.41	0.68	0.97	1.41	2.13	3.0	4.2	5.4	6.7	
$KMnO_4$	0	2.75	4.07	5.96	8.28	11.13	14.40	18.15					−44.0

contd.

Substance	H_2O	0	10	20	30	40	50	60	70	80	90	100	ΔH_∞
K_2MoO_4	0	63.8	64.25	64.50	64.7	65.0	65.2	65.6	65.9	66.3	66.5	66.8	
KNO_2	0	73.7	74.5	75.4	76.1	76.8	77.3	77.9	78.5	79.0	79.6	80.2	−15.5
KNO_3	0	11.7	17.7	24.0	31.3	39.0	46.0	52.2	57.8	62.8	67.0	71.0	−36.0
$KNaC_4H_4O_6$	3	24.25	32.0	40.1									
$K_2Ni(SO_4)_2$	6	3.26	4.31	5.61	7.17	8.97	11.02	13.33	15.88	18.68	21.74	25.04	
KOH	2	49.2	50.7	52.8	55.7	(32.5°) 58.3							+0.54
	1									61.7		65.1	+54.4
	0												
K_3PO_4	7	44.26	46.83	49.62	53.08	57.51	(45.4°) 63.80	64.08					
	3												
$K_2Pt(CN)_4$	5	10.5	16.7	(13.35°) 25.3	34.6	43.9	52.2	(52.40°) 58.2	62.1	(74.5°) 63.6	65.8		
	3												
	2												
	1												
KSCN	0	63.9		68.45	71.3	73.8	76.5	78.7	81.0	83.0	85.5	87.2	−25.5
K_2SO_3	0	47.52				50.37				53.15		55.53	+7.5
K_2SO_4	0	6.9	8.5	10.0	11.5	12.9	14.2	15.4	16.5	17.6	18.6	19.4	−26.4
$K_2S_2O_3$	2	49.0		60.8	63.7	(35°) 67.2	68.3	(56°) 70.4	71.8	(78.3°) 74.5	75.7		−18.8
	1												
	1/3												
	0												
$K_2S_2O_5$	0	22.1	26.5	30.8	34.8	39.0	42.5	46.0		51.9	52.9		−46.1
	2/3	21.7	27.4	33.2									−42.7
KSO_3NH_2	0		32.2	40.05	47.0	53.05							

191

K$_2$SeO$_4$	0	53.55	53.55	53.7	53.95	54.15	54.4	54.7	55.0		56.2		
K$_2$TiF$_6$	1	0.55	0.90	1.26									−69.1
La(NO$_3$)$_3$	6	50.03	52.3	54.65	57.6	62.30	65.35	71.1					
La$_2$(SO$_4$)$_3$	9	2.9			1.86	1.47					0.9		+16.7
LiBO$_2$	8 2	0.88	1.42	2.51	4.63	(36.9°) 7.40	7.48	8.43	9.48	10.58			
LiO · OCC$_6$H$_5$	1	27.9	29.2	30.8	32.4	33.7					34.8	35.6	
LiBr	3 2 1	58.8	(4°) 62.4	63.9	65.7	(40°) 67.2	67.9	69.1	69.5	71.0	71.9	72.7	+8.8 +26.4
Li$_2$CO$_3$	0	1.52	1.41	1.31	1.24	1.16	1.07	1.0	0.92	0.84	0.77	0.71	+15,5
LiCl	2 1 0	40.9	42.7	(18.5°) 45.3	46.3	47.3	48.3	49.6	51.1	52.8	54.8	(96°) 56.2	−3.8 +18.0 +36.4
LiI	3 1	60.2	61.1	62.2	63.1	64.2	65.2	66.9	69.7	(77°) 81.3	82.0	82.7	−1.1 +28.4
LiNO$_2$	1 1/2 0	43.95	45.27				62.6	(50.9°) 65.95	67.5	70.3	73.4	(94°) 76.6	
LiNO$_3$	3 1/2	34.6	37.8	42.3	(29.6°) 59.05	60.2	61.3	63.6	66.0				−33.1 −2.1
LiOH	1	10.64	10.8	10.99	11.27	11.68	12.12	12.76		14.21	16.05		+3.8
Li$_2$SO$_4$	1	26.3	25.7	25.2	25.25	25.1	24.3	24.45	24.2	24.1	23.4	23.5	+13.8
MgBr$_2$	6		49.8	50.3	51.1	51.6	52.3	53.0	53.4		55.8		+82.9
Mg(BrO$_3$)$_2$	6 2	42.3	45.5	48.6	51.4	54.5	57.3	60.8	64.6	(80°) 70.1	70.8	71.9	

contd.

Substance	H₂O	\multicolumn{11}{c}{t °C}	ΔH_∞
		0	10	20	30	40	50	60	70	80	90	100	
MgCl₂	6	34.6	34.9	35.3	35.8	36.5	37.2	37.9	38.8	39.8	41.0	42.3	+13.0
Mg(ClO₃)₂	6	53.27	54.5	57.2	60.6	63.7	(42°)						
Mg(ClO₄)₂	6	47.7	48.6	49.7	50.5	51.2	52.1						+6.3
MgCrO₄	7 5	32.06	33.87	(17.2°) 35.2	35.6	36.1	36.82	37.68	39.2				
MgI₂	8 6	54.7		58.3		63.4	(43.5°)			65.0	65.1	65.2	
Mg(IO₃)₂	10 4 0		5.87	(13.3°) 7.8	9.2	10.51	12.05	(57.5°) 13.2	13.3	13.4	13,5		−15.5
Mg(NO₃)₂	9 6 3	38.4	39.7	(10.5°) 41.2	(29.5°) 42.6	44.1		44.7		51.5	58.0		
MgSO₃	6 3	0.338	0.438	0.716	0.746	(40°) 0.95	0.844	0.76	0.69	0.62	0.61		
MgSO₄	7 6 1	18.0	22.0	25.2	28.0	30.8	(48°) 33.4	35.3	(69°)	35.8	34.6	33.5	−15.9 +2.3 +58.6
MgS₂O₃	6	30.69	31.9	33.3	34.7	35.9	37.1	38.7	39.0	42.8			
MgSiF₆	6	20.85		23.53		25.86	28.54	30.74					
MnCl₂	4 0	38.8	40.5	42.5	44.68	46.96	49.53	(58°) 52.06	52.52	52.98	53.2	53.5	+6.3 +67,0
MnF₂	4 0	0.8		0.97	(23.5°) 1.06	0.66		0.44				0.48	

Mn(NO$_3$)$_2$	6		52.8	57.0	(24.7°) 66.1		81.6	82.1	82.9			−25.5
	4											+54.0
	1											−14.7
	0											+0.17
MnSO$_4$	7	34.6	(8.6°) 37.4	38.6	(23.9°) 39.35	37.5	36.3	34.9	34.2	31.3	29.0	26.0
	5											
	1											
MoO$_3$	2			0.15	0.26	0.33	0.40	0.42	(70.5°) 0.46	0.51		
	1											
(NH$_4$)$_2$B$_4$O$_7$	4	3.75	5.26	7.63	10.80	15.77		27.0			52.68	
NH$_4$OOCC$_6$H$_5$	0		15.4	17.5	19.6							
NH$_4$Br	0	37.3	40.0	42.6	45.0	47.3	49.4	51.2	52.8	54.4	56.0	57.4	−18.8
(NH$_4$)$_2$C$_4$H$_4$O$_6$	0	31.1	35.5	38.6	41.3	43.3	44.9	46.5				
(NH$_4$)$_2$C$_2$O$_4$	1	2.31	3.11	4.26	5.73	7.56	9.73	12.25	15.10	18.30	21.84	25.73	−48.1
NH$_4$Cl	0	22.7	24.9	27.1	29.3	31.4	33.5	35.6	37.6	39.6	41.6	43.6	−15.9
NH$_4$ClO$_4$	0	10.92		19.0		27.1		33.7		40.4		46.9	−26.8
(NH$_4$)$_2$CrO$_4$	0	19.78		25.35	28.80		34.40		37.21				−24.3
(NH$_4$)$_2$Cr$_2$O$_7$	0	15.37		26.23	31.74	36.91	42.03	46.24	44.0	53.49		60.89	−54.0
NH$_4$F	0	41.81	42.55	45.28	47.05	52.01		52.62	54.05				−6.3
NH$_4$OOCH	0	50.8		58.9		67.1		75.7		84.2			
NH$_4$H$_2$AsO$_4$	0	25.2		32.7		39.0		45.3		51.7	55.0		
NH$_4$HCO$_3$	0	10.6	13.9	17.8	22.1	26.8	31.6	37.2	44.0	52.2	63.0	78.0	−28.1
NH$_4$HF$_2$	0	24.85	31.96	37.56		50.05		61.0		74.53		85.55	
NH$_4$H$_2$PO$_4$	0	18.5	22.8	27.2	31.7	36.3	40.9	45.2	50.0	54.2	58.9	63.4	−18.0

contd.

Substance	H_2O	\multicolumn{11}{c	}{t °C}	ΔH_∞
		0	10	20	30	40	50	60	70	80	90	100	
NH_4HSO_3	0	72.0		77.0		80.4		85.4					
NH_4I	0	60.7	62.1	63.3	64.5	65.7	66.6	67.7	68.9	69.6		71.3	−15.1
$(NH_4)_2Mg(SO_4)_2$	6	10.58	12.75	15.23	17.84	20.51	23.18	26.02		32.58		39.66	
NH_4NO_2	0	55.33	60.03	64.9	72.5								−20.1
NH_4NO_3	0	54.2	60.0	65.5	70.4	74.6	77.8	80.7	83.3	85.7	88.2	90.3	−27.2
$(NH_4)_2PtCl_6$	0	0.29	0.37	0.50	0.63	0.80	1.02	1.14	1.74	2.16	2.58	3.25	−35.2
NH_4SCN	0		58.23	62.54	66.26	70.1	74.1	77.2	81.0				−23.9
$(NH_4)_2SO_3$	1 0	32.40	35.05	37.80	40.77	43.96	47.26	50.94	54.71	58.99	(80.8°) 60.00	60.44	−18.0 −6.3
$(NH_4)_2SO_4$	0	41.35	42.05	42.85	43.75	44.7	45.8	46.64	47.54	48.47	49.44	50.42	−11.3
$(NH_4)_2S_2O_6$	1/2	57.05	60.14	62.43	64.10								
$(NH_4)_2S_2O_8$	0	37.0	40.1	43.4		50.9		56.1		62.0			−38.5
$(NH_4)_2S_3O_6$	0	53.2		56.4	58.4								
$(NH_4)_2S_4O_6$	0	51.2		54.3	56.2								
$NH_4SO_3NH_2$	0	57.6	62.5	66.7	70.0	73.7	78.1		84.1				
$(NH_4)_2SeO_4$	0			54.1	55.12								
$(NH_4)_2SiF_6$	0	10.94	13.95	17.16	20.2	23.2	26.15	28.75				37.90	−35.2
NH_4VO_3	0			0.48	0.92	1.32	1.61	2.49	3.04				

Compound	n													Heat
NaOOCCH₃	3 0	26.6	28.9	31.7	35.2	39.5	45.3	58.2		60.5		63.0		−20.5 +16.7
Na₂B₄O₇	10 4	1.18	1.76	2.58	3.85	6.00	9.55	(58.5°) 14.82	17.12	19.88	23.31	28.22		−108.0
NaOOCC₆H₅	0	38.5		38.6		38.7		39.2	39.9	40.6	41.65	42.7		
NaBr	2 0	44.47	46.0	47.6	49.6	51.5	53.8	(51°) 54.1	54.3	39.9	40.6	41.65	42.7	−19.3 −2.5
NaBrO₃	0	21.5	23.24	26.69	29.85	32.80	35.55	38.5		43.1		47.6		
NaCN	2 0		32.5	37.0	41.6	(35°) 45.1	45.2							−19.7 −2.1
Na₂CO₃	10 7 1	6.54	10.8	18.1	28.2	(32.0°) (35.7°) 32.8	32.2	31.6	31.3	31.1	31.0	30.9		−67.9 −42.6 +10.5
Na₂C₂O₄	0	2.67	3.01	3.39	3.76	4.04	4.34	4.60	4.91	5.24	5.55	5.87		−18.8
NaCl	2 0	26.30 (0.1°) 26.31	26.7	26.38	26.50	26.65	26.83	27.05	27.27	27.54	27.80	28.2		−5.0
NaClO	5 2.5	22.7	26.7	34.8	(23.0°) 50.0	52.5	56.5	55.0						
NaClO₂	3 0		36.0	40.9	46.4	(37.4°) 52.2	53.6							
NaClO₃	0	44.3	46.7	48.9	51.2	53.4	55.6	57.9	60.1	62.4	64.6	66.8		−22.6
NaClO₄	1 0	62.64		68.71	70.88	73.16	(50.8°) 74.3					76.75		−23.0 −15.1
Na₂CrO₄	10 6 4 0	24.2	32.1	(19.5°) 44.3	(25.9°) 46.8	48.9	51.0	53.5	(65°) 55.2	55.5	55.8	56.1		−66.2 −31.4 +10.0

contd.

Substance	H$_2$O	\multicolumn{11}{c	}{t °C}	ΔH_∞
		0	10	20	30	40	50	60	70	80	90	100	
Na$_2$Cr$_2$O$_7$	2 0	62.00	63.0	64.3	66.3	68.3	70.5	72.9		79.0	(82.5°) 80.2	80.6	
NaF	0	3.53	3.71	3.90	4.05	4.21	4.35	4.47	4.56	4.66	4.75	4.83	−1.3
Na$_4$Fe(CN)$_6$	10 0		12.8	15.8	19.3	22.9	26.3	30.4	34.16	38.3	(81.7°) 39	39	
NaOOCH	3 2 0	30.47	38.5	(15.3°) 46.2	(24.5°) 50.6	52.0	53.4	55.0		57.6		61.4	−1.7
Na$_2$HAsO$_4$	12 7 5 1 0	5.59	11.52	25.31	(20.5°) 33.0	41.03	49.8	(56.2°) 58.8	(67.4°) 64.8	65.1	65.3	(99.5°) 66.5	
NaHCO$_3$	0	6.48	7.00	8.72	9.96	11.29	12.64	14.10	15.3	16.9	18.0	19.1	−17.2
NaHC$_2$O$_4$	1	0.92	1.20	1.60	2.28	3.43	4.70	6.42	8.58	11.0	13.4	15.9	−38.9
NaH$_2$PO$_3$	2.5	25.1	29.4	34.2	39.3	44.8							−22.2
Na$_2$HPO$_3$	5.5	80.72	80.9	81.06	84.96	92.2							−18.8
NaH$_2$PO$_4$	2 1 0	36.5	41.0	45.5	51.5	57.0	(41°) 61.5	(55°) 65.0	66.0	68.0	69.5	71.0	+23.4
Na$_2$HPO$_4$	12 7 2 0	1.6	3.5	7.1	17.8	(35.4°) 34.5	(48.1°) 44.5	45.3	46.5	48.3	50.1	(95.1°) 50.8	−95.0 −48.6 −1.7 +21.8

Na₂H₂P₂O₇	6	4.28	6.50	10.70	(27°) 14.58	15.02					−58.6
	0										
Na₃HP₂O₇	9			13.9	23.8	(32.9°) 19.9	18.8				
	1										
NaI	2	61.54	62.8	64.1	65.6	67.2	69.4	72.0	(68.1°) 74.6	74.7	74.9	75.1	−16.3
	0										+6.3
NaIO₃	5	2.42	4.39	(19.85°) 9.63	11.71	14.0	16.5	19.0	(73.4°) 21.0	22.8	24.8	
	1										
	0										
NaIO₄	3		5.4	9.3	16.6	(34.5°) 23.3	27.4				
	0										
Na₂MoO₄	10	30.63	(10°) 39.28	39.30	39.75					45.57	
	2										
NaNO₂	0	41.65	43.4	45.2	46.7	48.8	50.8	53.0	55.1	57.0	59.3	61.6	−15.1
NaNO₃	0	42.2	44.6	46.8	49.0	51.2	55.4			59.7		63.7	−21.3
NaOH	4	29.6	(5°) 34.0	(12°) 52.2	54.3	56.3	59.2	63.5	(61.8°)	75.8	76.7	77.7	+21.3
	3.5										+42.7
	1										
	0										
Na₃PO₄	12	4.2	7.5	10.7	14.0	17.0	22.7	(55°) 28.4 (70°)	32.6 32.6	(70°) 37.5	40.4	43.5	−62.8
	8										
	6										
Na₄P₂O₇	10	2.23		5.22	7.04		13.98	19.75	27.49	(79.5°) 35.00	32.7		−50.2
	0										+50.2
Na₂PtCl₆	6	38.7			47.87	50.67	54.07	57.07	59.12	65.25	68.10		−44.4
Na₂S	9		13.36	15.8	18.4	22.1	(48.9°) 28.48	29.92	31.38	33.95		37.20	−69.9
	5.5										−27.2

contd.

| Substance | H$_2$O | \multicolumn{11}{c}{t °C} | ΔH_∞ |
		0	10	20	30	40	50	60	70	80	90	100	
NaSCN	1 0			57.5	62.7	(30.4°) 63.64	64.3	65.05	65.86	66.7	67.9	69.12	−7.5
Na$_2$SO$_3$	7 0	12.5	16.1	21.3	25.9	(32.95°) 26.3	25.7	24.0	23.8	21.8		21.6	−46.9 +11.3
Na$_2$SO$_4$	10 0	4.3	8.2	15.9	29.1	(32.4°) 32.4	31.7	31.1	30.6	30.1	29.8	29.6	−78.5 +1.2
Na$_2$S$_2$O$_3$	5 2 0	33.40	37.37	41.20	45.4	50.4 59.38	(48.17°) 62.28	65.68	(66.5°) 69.05	69.86			−47.7 +7.5
Na$_2$S$_2$O$_4$	2			18.03									
Na$_2$S$_2$O$_5$	7 0	31.1	(5.5) 38.7	39.6	40.6	41.6	43.0	44.38	45.5	46.9	48.3		
Na$_2$S$_2$O$_6$	6 2	5.9	(9.1°) 10.00	13.1	16.40	19.8	23.2	26.5	29.8	33.0	36.00	39.2	−49.0
Na$_2$SO$_3$NH$_2$	1 0	44.7		53.1	57.6	(38.3°) 62.8	64.8						
Na$_2$SeO$_4$	10 0	11.74			44.05	(31.8°) 45.3	44.5					42.14	
Na$_2$SiF$_6$	0	0.41	0.53	0.67	0.84	1.02	1.22	1.44	1.68	1.92	2.17	2.42	
Na$_2$Sn(OH)$_6$	1 0	33.2 31.5			37.0 29.8	38.3 28.0	41.1						
NaWO$_4$	10 2	41.7	(6°)	42.1						47.6		49.31	
Nd$_2$(SO$_4$)$_3$	15 8	11.5	(10°) 8.8	6.6	5.0	3.9	3.2	2.7	2.4		1.1	1.2	

199

		53.0	55.0	56.7	58.0	59.1	60.0	60.4	60.5	60.6	60.7	60.8	
NiBr$_2$	6	53.0	55.0	56.7	58.0	59.1							−4.8
NiCl$_2$	6 4 2	34.8		38.3	(28.8°) 41.7	42.4	43.2	44.8	(64.3°) 46.2	45.9		46.7	+43.1
Ni(ClO$_4$)$_2$	5	51.2	51.8	52.5	53.2	54.0							
NiF$_2$	4		2.49	2.50	2.50	2.50	2.50	2.50	2.51	2.51	2.52	2.52	
NiI$_2$	6 4	55.4	57.5	59.7	61.7	63.5	64.7	64.8	65.0	65.2	65.3		
Ni(NO$_3$)$_2$	6 4 2	44.2		48.5	51.3	54.3	58.2	(54°) 61.2		65.5	(85.4°) 68.1	69.2	−31.4
NiSO$_4$	7 6	21.6	24.6	27.5	30.4	(31.2°) 32.3	34.3	36.0	37.7	39.7	41.8	43.8	−18.0
OsO$_4$	0	5.0	5.44	6.04			67.6						
Pb(OCCH$_3$)$_2$	3 0	16.5	22.8	30.7	41.1	53.7							−24.7 +5.9
PbBr$_2$	0	0.44		0.83	1.47	1.8	2.2			3.19		4.53	
PbCl$_2$	0	0.66		1.08	1.17		1.75			2.55		3.15	−26.0
PbI$_2$	0	0.04		0.06	0.09		0.17			0.29		0.42	
Pb(NO$_3$)$_2$	0	28.3	32.5	36.1	39.7	42.8	45.9	48.7		53.6		58.2	−31.8
PbSiF$_6$	4 2	65.48		68.97			74.16	(60°) 80.11		81.06		82.25	
Pr(NO$_3$)$_3$	6			59.9	61.9	64.4	70.4	(56°)?					
Pr$_2$(SO$_4$)$_3$	8 5	16.5	13.5	11.2	9.0	7.2	6.0	4.8	4.0	3.4	(90°) 1.1	0.9	

contd.

Substance	H₂O	\multicolumn{11}{c}{$t\ °C$}	ΔH_∞
		0	10	20	30	40	50	60	70	80	90	100	
PtCl₄	8	39.99				62.38	(40°)	74.00	(60°)	(80°)			−7.5
	5					62.38	69.03	74.00	74.66	78.6			
	4										82.5		
	3												
PtK₂Cl₆	0	0.73	0.89	1.10	1.39	1.72	2.06	2.57	3.09	3.65	4.26	4.92	
RaBr₂	0			41.4									
RaCl₂	0			19.7									
Ra(NO₃)₂	0			12.2									
RbOOCCH₃	0	83.5				85.9	86.5				88.7	89.3	
RbAl(SO₄)₂	12				2.14	3.11							
RbBr	0	47.2	50.2	53.0	54.7			61.3				65.6	−25.5
RbCl	0	43.5	45.8	47.7	49.4	50.9	52.2	53.6	54.8	56.0	57.1	58.9	−18.4
RbClO₃	0	2.09	3.0	5.12	7.4		13.7					38.6	−47.7
RbClO₄	0	0.49	0.59	0.99		2.2	3.3	4.6	6.2	8.4	11.2	15.3	−56.9
RbOOCH	1		81.3	(16.5°)			89.23	(51.0°)	90.9	91.82	92.7	93.8	
	1/2							90.06					
	0												
RbI	0	55.5	58.95		63.8		68.7		71.3		72.6	73.7	−27.6
RbNO₃	0	16.3	24.8	34.6	44.8	53.9	60.9	66.7	71.5	75.6	78.9	81.9	−37.7
Rb₂SO₄	0	27.3	29.9	32.5	34.9	36.9	38.7	40.3	41.7	42.9	44.0	45.0	−28.1
SbCl₃	0	85.7	90.18	91.41	93.16	95.03	97.77						

Formula	n	0°	10°	20°	30°	40°	50°	60°	70°	80°	90°	100°	q
SbF_3	0	79.4	81.6	84.9									−6.7
SeO_2	1	65.8	69.1	71.9	74.5	77.0	79.2	81.5					
SnI_2	0			0.97	1.14	1.38	1.66	2.03	2.41	2.86	3.34	3.87	−24.3
$Sr(OOCCH_3)_2$	4	26.9	(8.4°) 30.04		28.26		27.19						+24.7
	1/2												−26.8
$SrBr_2$	6	46.0	48.3	50.6	52.8	55.2	57.6	60.0		26.6	26.5	26.6	69.0
$SrCl_2$	6	30.3	32.3	34.6	37.0	39.5	42.0	45.0	(70°) 46.2	47.5		50.4	−31.4
	2												+10.5
SrI_2	6	62.3	64.0	65.7		68.5		73.0	(84°) 78.5	79.3			−19.7
	2												
$Sr(NO_2)_2$	1	34.57	37.06	39.5	44.07	45.0	46.02	49.2	51.6	54.52	56.1		
$Sr(NO_3)_2$	4	28.2	35.0	40.7	(29.3°) 47.1	45.0	47.7	48.3	48.8	49.2	49.6	51.0	−51.9
	0												−20.1
$Sr(OH)_2$	8				1.02	1.48	2.20	3.29	5.00	8.38	(85°) 11.61	10.83	−59.9
	1												+22.2
SrS_2O_3	3	8.78	12.73	17.4	22.1	26.80							
$Th(NO_3)_4$	6	65.0	65.2	65.6	66.4						76.3	78.6	
$Th(SO_4)_2$	9	0.73	0.97	1.36	1.95	2.90	4.96		1.07				−31.4
	8	1.0	1.23	1.52	2.34	3.23		6.05					
	6	1.47	1.52	1.95	2.39		2.47	1.60					+20.9
	4					3.88							
$TlBrO_3$	0			0.3	0.5	0.75							
Tl_2CO_3	0			5.3	6.8	8.3	9.7	11.1				21.4	
$Tl_2C_2O_4$	0		1.8										
$TlCl$	0	0.17	0.24	0.33	0.44	0.59	0.8	1.0		1.5		2.2	−43.1

contd.

Substance	H$_2$O	\multicolumn{11}{c}{t °C}	ΔH_∞
		0	10	20	30	40	50	60	70	80	90	100	
TlClO$_3$	0	1.96		3.77			11.24			26.82		36.43	
TlClO$_4$	0	5.6	7.4		16.5		28.4		39.5	44.9			
TlNO$_2$	0	15.15	22.44	28.75	34.73	45.53	55.10	68.40	86.1	91.6	94.4	95.8	
TlNO$_3$	0	3.76	5.86	8.72	12.51	17.33	23.33	31.55	41.01	52.6	66.6	80.54	−41.8
TlOH	0	20.66	23.99	26.52	29.51	34.13							−13.4
Tl$_3$PO$_4$	0			0.5								6.71	
TlSCN	0			0.3	0.5	0.7							
Tl$_2$SO$_4$	0	2.63	3.57	4.64	5.80		8.44	8.89	11.31	12.77	14.19		−33.5
UO$_2$C$_2$O$_4$	3			0.50			1.00					3.06	
UO$_2$(NO$_3$)$_2$	6	49.7		54.4	57.9	62.3		(58.6°)	77.25		80.90	82.57	−23.0
	3												+8.0
YCl$_3$	6	42.36	42.65	42.87	43.07	43.30	43.45	43.60	43.75	43.86			
Y(NO$_3$)$_3$	4	48.3			59.6			66.8					
Y$_2$(SO$_4$)$_3$	8		7.14	6.94		3.61	3.09			2.8	2.24		
Yb$_2$(SO$_4$)$_3$	8	30.65		22.0		15.7		9.4		6.4		4.5	
Zn(OOCC$_6$H$_5$)$_2$	0			2.46	2.20	1.4							
ZnBr$_2$	2	79.55			84.08	(35°) 85.53		86.08		86.57		87.05	+62.8
	0												

ZnCl₂	2.5	70.1	73.1	(11.5°)							+69.5
	1.5		76.8	78.6						86.0	
	1			79.8			(28°)	83.0	84.4		
	0						81.9				
ZnI₂	2	81.16	82.06	83.0			81.66	82.37	83.05	83.62	+48.6
	0	81.11									
Zn(NO₃)₂	6	48.0			58.2	(34°)		(37°)			−24.7
	4					67.9		81.5			
	2					79.4		(52°)			
	1							87.7		89.9	
ZnSO₄	7	28.6				38.1	(37.9°)	(48.8°)	40.0	37.7	−18.0
	6						41.2	42.8	38.8		−3.5
	1							41.9			+41.9
ZnSiF₆	6	33.73		35.16		37.04		38.49	40.95	42.18	
acetamide	0	58.0	63.6	69.7	75.6	81.5	89.5				−8.4
acetanilide	0		0.48	0.52	0.62	0.86		2.1	4.6		
adipic acid	0	0.8	1.0	1.8	2.9	4.8		15.2	41.2	61.5	
aniline	0		3.2	3.3		3.8		4.5	5.7	7.2	
benzoic acid	1	49.0	54.1	59.3	64.7	0.55	0.77	1.14	2.64	5.55	−27.2
	0	0.17	0.21	0.29	0.41						
citric acid	0					68.3		73.5	78.8	84.0	−22.6
phenol	0		7.7	7.9	8.5	9.5	11.7	15.8	(66.5°)		
glucose	1	31.5	41.2	47.9	54.5	61.5		73.7	81.5		
glycine	0	12.4	15.3	18.4	21.3	24.8		31.0	36.3	41.2	
hydroquinone	0	3.8	5.1	6.7	8.8	11.5		25.9	46.8	66.4	

contd.

Substance	H$_2$O	\multicolumn{11}{c	}{t °C}	ΔH_∞
		0	10	20	30	40	50	60	70	80	90	100	
lactose	1	10.9	13.0	16.3	20.1	25.0		36.5		50.5		60.5	+10.5
succinic acid	0	2.7	4.3	6.3	9.5	13.9	19.6	26.4		41.5		54.7	
urea	0	40.1	44.4	51.2	57.4	62.1		71.4		80.0		88.0	−14.2
oxalic acid	2	3.42	5.73	8.69	12.5	17.7	23.9	30.7		45.8		54.5	−35.6
pentaerythritol	0	3.8	4.8	5.7	7.4	11.5		18.0		28.6		50.0	
pyrocatechol	0			31.1		63.2		80.5		91.8		98.8	−14.7
resorcinol	0	39.8	45.9	55.2	63.0	69.2		79.6		86.4		91.4	−16.3
salicylic acid	0	0.13	0.15	0.20	0.27	0.41		0.90		2.2		7.5	−26.4
sucrose	0	64.2	65.6	67.1	68.7	70.4	72.3	74.2		78.4		83.0	−5.4
sulphanilic acid	0	0.45	0.80	1.1		2.0		2.9		4.3		6.3	
tartaric acid	0	53.5	55.8	58.2	61.0	63.8	66.1	68.6		73.2		77.5	−14.7

9. TABLES OF SOLUBILITIES IN TERNARY SYSTEMS

The following data are given in the tables, which characterize equilibria in three-component systems using the method described in Section 5.2.4.

Component (1) represents the solid equilibrium phase. In the second column is given the number of molecules of water of crystallisation in the equilibrium solid phase.

Component (2) is another substance contained in the three-component solution; water is the third component.

Q_1, Q_2 and Q_3 are the constants in eqn. 5.195 or 5.202, expressing the effect of the other component on the relative activity coefficient of the first component. In the column headed "data for c_2" the concentration range is defined for which the validity of constants Q_1, Q_2 and Q_3 was verified. The series is usually terminated by the eutonic point.

The compounds are ordered alphabetically.

Component 1	H_2O	Component 2	t °C	Q_1	Q_2	Q_3	Data for c_2
$AgCH_3COO$	0	$NaClO_4$	25	−0.107	0	0	3.4259
	0	$AgClO_4$	25	0.36	1.0	−0.5	0.05−0.30
	0	$Ca(CH_3COO)_2$	25	0.82	−0.38	0	0.078−0.79
AgCN	0	KCN	25	0.30	0.02	0	0.03−0.26
$AgNO_3$	0	$Pb(NO_3)_2$	25	0.053	−0.012	0	0.77−2.20
	0	$Ba(NO_3)_2$	30	0.056	0	0	0.10−0.38
	0	KNO_3	30	0.043	−0.002	0	2.39−5.61
	0	NH_4NO_3	30	0.019	−0.0003	0	3.38−16.53
Ag_2SO_4	0	Li_2SO_4	25	0.634	−0.12	0	0.5−3.12

contd.

Component 1	H$_2$O	Component 2	t °C	Q_1	Q_2	Q_3	Data for c_2
BaBr$_2$	2	Ba(ClO$_3$)$_2$	10	0.0054	0	0	0.2859
	2		25	0.0089	0	0	0.3906
	2	CdBr$_2$	25	0.102	−0.008	0	0.59−3.63
	2		30	0.090*	−0.004	0	0.73−3.75
	2		35	0.090	−0.004	0	0.53−4.23
	2		50	0.079	−0.003	0	0.31−4.47
	2		75	0.076	−0.003	0	0.65−5.56
	2		100	0.069	−0.005	0	0.6−1.2
	2	HBr	0	−0.055	0	0	0.2−9.72
	2		25	−0.050	0	0	0.28−2.7
	0	HgBr$_2$	4.5	0.084	−0.0030	0	0.61−7.19
	0		10.4	0.084	−0.0030	0	0.73−8.35
	0		25	0.084	−0.004	0	0.31−9.9
Ba(BrO$_3$)$_2$	2	Ba(NO$_3$)$_2$	10	1.2	−1.14	0.386	0.33−1.39
	2		45	0.6	−0.33	0.061	0.65−1.65
BaCl$_2$	2	Ba(BrO$_3$)$_2$	10	0.1711	0	0	0.0049
	2		25	0.1418	0	0	0.0079
	2		45	−0.0205	0	0	0.0142
	2	Ba(ClO$_3$)$_2$	20	0.025	0	0	0.25−1.24
	2		25	0.004	0	0	0.24−0.75
	2	Ba(NO$_3$)$_2$	20	0.110	0	0	0.06−0.39
	0		20	−0.0009	0	0	0.05−0.33
	0		30	0.107	−0.01	0	0.08−0.45
	0		40	−0.10	0.09	0	0.33−0.42
	0		60	−0.12	0.08	0	0.33−0.42
	2	Ba(OH)$_2$	25	0.086	0	0	0.07−0.15
	2		30	0.115	0	0	0.1452
	2	BeCl$_2$	25	−0.187	−0.075	0	0.77−3.36
	2	CdCl$_2$	25	0.140	−0.018	0	0.43−1.21
	2		30	0.140	−0.017	0	0.51−1.31
	2		35	0.135	−0.018	0	0.39−0.92
	2		50	0.125	−0.018	0	0.43−1.86
	2		75	0.11	−0.006	0	0.9−2.74
	2		100	0.11	−0.008	0	0.3−4.36
	2	CoCl$_2$	20	−0.25	0	0	0.63−4.06
	2	CuCl$_2$	30	−0.08	0	0	0.52−5.73
	2	LiCl	25	0.0025	0	0	0.48−10.01
	2	MgCl$_2$	50	−0.10	0	0	1.65−5.07
	2	NH$_4$Cl	25	0.017	−0.0011	0	1.47−7.32
	2	NaCl	20	−0.009	−0.003	0	1.77−5.81
	2		30	−0.009	−0.0025	0	0.64−5.61

contd.

Component 1	H$_2$O	Component 2	t °C	Q_1	Q_2	Q_3	Data for c_2
Ba(ClO$_3$)$_2$	1	BaBr$_2$	10	−0.0424	0	0	3.1182
	1		25	−0.0544	0	0	3.2761
	1	BaCl$_2$	25	−0.033	0	0	0.43−1.57
	1	NaClO$_3$	25	−0.016	0.004	0	3.4−7.5
Ba(NO$_3$)$_2$	0	HNO$_3$	25	0.020	0	0	4.15−20.7
	0		30	−0.133	0.073	−0.0030	0.48−2.32
	0	BaCl$_2$	20	0.18	−0.020	0	0.42−1.7
	0		30	0.14	0	0	0.34−1.89
	0		40	0.083	0	0	0.25−1.87
	0		60	0.048	0	0	0.25−2.09
	0	KNO$_3$	21	0.184	0	0	0.23−1.62
	0		25	0.178	−0.009	0	0.42−1.86
	0		35	0.153	−0.0075	0	1.62−2.71
	0	NH$_4$NO$_3$	30	0.080	−0.0017	0	1.46−28.6
	0	NaNO$_3$	0	0.173	−0.0095	0	0.05−8.34
	0		20	0.080	0	0	0.73−7.25
	0		30	0.060	0.005	0	0.2−7.0
	1	TlNO$_2$	25	0.09	0	0	0.23−0.42
Ba(OH)$_2$	8	Ba(SCN)$_2$	25	0.25	−0.034	0.0040	0.77−1.49
	8	BaCl$_2$	25	0.36	−0.115	0	0.19−1.21
	8	Ba(ClO$_3$)$_2$	25	0.35	−0.100	0	0.33−1.25
Ba(SCN)$_2$	3	AgSCN	25	0.062	−0.004	0	0.7−6.56
	3	Ba(OH)$_2$	25	0.057	−0.0015	0	0.33−0.52
	3	KSCN	25	0.035	−0.0017	0	4.98−9.79
	3	NH$_4$SCN	25	0.026	−0.0003	0	7.0−14.97
	3	NaSCN	25	−0.0016	0.0004	0	4.29−11.31
BaS$_2$O$_6$	2	MgS$_2$O$_6$	20	−0.026	−0.02	0	0.45−2.82
	2		30	−0.054	−0.023	0	0.48−2.90
BeCl$_2$	4	HCl	0	−0.0014	0	0	0.2−6.29
	4		20	−0.005	0	0	0.38−5.09
	4		30	−0.012	0	0	0.31−2.31
	4	LiCl	0	−0.003	0	0	2.26−5.59
	4		30	−0.008	0	0	0.85−2.53
	4		40	0.009	0	0	0.91
Be(NO$_3$)$_2$	4	HNO$_3$	0	0.005	0	0	11.8−65.4
	4		20	0.0095	0	0	9.11−41.03

contd.

Component 1	H$_2$O	Component 2	t °C	Q_1	Q_2	Q_3	Data for c_2
BeSO$_4$	4	CuSO$_4$	30	0	−0.09	0	0.10−0.47
	4		80	0.016	−0.014	0	0.1−1.79
	4	FeSO$_4$	25	−0.19	0.23	0	0.33−0.78
	4		60	−0.19	0.07	0	0.65−1.74
	4	H$_2$SO$_4$	25	−0.019	0	0	1.9−12.97
	2		50	−0.02	0.0017	0	2.59−8.97
	2		75	−0.01	0.0017	0	0.35−4.09
	1		100	−0.033	0.0020	0	2.0−11.4
	4	K$_2$SO$_4$	0	0	−0.05	0	0.08−0.12
	4		25	0	−0.04	0	0.12−0.16
	4		50	0	0.075	0	0.06−0.19
	4		75	0	0	0	0.25−0.33
	4	MnSO$_4$	25	0.16	−0.012	0	1.02−2.29
	4	(NH$_4$)$_2$SO$_4$	0	0.101	−0.01	0	0 19−4.04
	4		25	0.101	−0.008	0	0 76−3.66
	2		25	0.060	0	0	0.66−4.71
	2		50	0.117	−0.01	0	1.8−3.35
	2		60	0.117	−0.012	0	0.41−2.98
	2		75	0.117	−0.012	0	0.77−2.39
	2		99.5	0.06	0	0	1.73−2.21
	4	Na$_2$SO$_4$	0	0	0.3	0	0.03−0.29
	4		25	0.028	0	0	0.44−2.33
	4		50	0.043	0	0	0.77−1.99
	4		75	0.052	0	0	0.5−1.85
	4		86	0.056	0	0	0.33−1.99
	4	ZnSO$_4$	0	−0.043	0.0020	0	0.17−1.09
	4		25	−0.043	0.004	0	0.44−2.12
	4		50	−0.043	0.009	0	0.2−2.46
	2		99	−0.009	0	0	0.2−0.7
CaBr$_2$	0	HBr	25	−0.030	0.009	0	1.0−3.3
CaCr$_2$O$_7$	0	K$_2$Cr$_2$O$_7$	50	0.014	0.07	0	0.11−0.58
CaCl$_2$	6	Ca(ClO)$_2$	0	0.015	0	0	0.18−0.29
	6	Ca(ClO$_3$)$_2$	20	0	0.01	0	0.6−2.55
	4		20	−0.03	0.01	0	2.55−4.23
	6		25	0.052	0.03	0	0.2−0.88
	4		25	−0.03	0.009	0	0.88−3.07
	2		75	−0.004	0	0	0.97−8.43
	6	Ca(NO$_3$)$_2$	25	0.07	0	0	0.34−0.97
	4		25	0.008	0	0	0.97−3.0
	4		30	0.020	0	0	1.49−2.55
	6	CoCl$_2$	0	0.034	0	0	0.2−0.35
	6		25	0.034	0	0	0.39−1.62
	2		50	0.003	0	0	1.24−2.42

contd.

Component 1	H_2O	Component 2	t °C	Q_1	Q_2	Q_3	Data for c_2
$CaCl_2$	6	HCl	0	−0.016	0	0	1.8−17.5
	6		25	0	0.014	0	0.96−1.75
	4		25	−0.005	0	0	1.7−11.7
	2		25	−0.010	0	0	11.7−11.88
	6	$HgCl_2$	25	0.049	0	0	1.44−2.21
	6	KCl	0	0.042	0	0	0.14−0.43
	6		20	0.042	0.91	0	0.15
	6		25	0.042	0	0	0.78−0.84
	4		30	0.032	0	0	0.08−1.27
	4		35	0.034	0	0	0.22−2.01
	2		50	0.030	0	0	1.3−1.7
	4	LiCl	25	−0.013	0	0	1.26−3.73
	2		25	−0.040	0.0015	0	5.4−8.83
	4		40	−0.013	0.010	0	0.89−1.35
	2		40	−0.013	0.001	0	1.35−15.6
	6	$MgCl_2$	−45	−0.062	−0.006	0	0.29−0.99
	6		−40	−0.062	0.001	0	0.30−1.36
	6		−30	−0.062	−0.009	0	0.33−1.33
	6		−25	−0.042	−0.01	0	0.3−2.75
	6		−20	−0.042	−0.01	0	0.3−2.75
	6		−15	−0.042	−0.010	0	0.3−1.9
	6		−10	−0.042	−0.0092	0	2.2−25.4
	6		−5	−0.036	−0.0097	0	0.3−4.3
	6		0	−0.06	0	0	0.43−4.01
	6		15	0	−0.065	0	0.50−3.27
	6		25	0	−0.05	0	0.78−1.25
	4		25	0.008	−0.022	0	0.56−1.90
	4		35	0	−0.030	0	0.38−1.33
	2		75	0	−0.022	0	0.63−1.31
	2		110	−0.015	−0.025	0	1.07−1.24
	2	$ZnCl_2$	60	−0.003	0.001	0	1.0−17
$Ca(ClO_3)_2$	2	$CaCl_2$	20	−0.032	0.0005	0	1.89−6.86
	2		25	−0.032	0.0012	0	1.48−7.65
	2		55	−0.016	0.0012	0	4.3−6.97
	0		75	−0.032	0.0015	0	0.9−12.77
$Ca(ClO)_2$	3	$CaCl_2$	0	−0.07	0	0	0.3−5.37
$Ca(NO_3)_2$	4	$CaCl_2$	25	−0.075	0.022	0	0.47−4.62
	4		30	−0.053	0.022	0	1.02−2.51
	4	HNO_3	0	−0.013	0.0005	0	6.0−32.7
	2		0	0.0008	0	0	32.7−54.1
	4		20	0.001	0	0	9.69−21.58
	4		25	0.001	0	0	1.26−16.5
	4		50	−0.0045	0	0	0.39−26.9
	4	KNO_3	0	0.045	0	0	0.6−3.9
	4		20	0.032	−0.0017	0	0.8−6.4
	4		25	0.024	0	0	0.6−5.9
	4		30	0.029	0.0016	0	3.11−6.41

contd.

Component 1	H_2O	Component 2	t °C	Q_1	Q_2	Q_3	Data for c_2
$Ca(NO_3)_2$	4	NH_4NO_3	0	0.033	−0.0003	0	1.8−12.4
	4		10	0.033	−0.0007	0	2.08−12.21
	4		20	0.033	0	0	1.05−4.69
	4		25	0.033	0	0	2.65−5.4
	4		30	0.033	0	0	2.6−6.24
	4		40	0.033	0	0	0.1−0.6
	4	$NaNO_3$	0	0.024	0	0	0.5−3.16
	4		20	0.021	0	0	0.5−3.41
	4		25	0.023	0	0	2.2−4.03
	3		50	0.021	0	0	6.69
	0		94.5	0.015	−0.0003	0	2.4−13.8
$Ca(SCN)_2$	4	AgSCN	20	0.036	0	0	0.66−1.46
	4		25	0.047	−0.01	0	0.65−1.27
	4	NaSCN	25	0.021	−0.002	0	1.0−3.41
CaS_2O_3	6	$CaCl_2$	25	−0.068	0.006	0	0.83−5.07
	6	$Na_2S_2O_3$	9	0	0	0	1.05−2.7
	6		25	0.02	0	0	0.95−3.85
$CaSeO_4$	2	K_2SeO_4	25	0.50	−0.20	0.0120	0.77−2.25
	2	$(NH_4)_2SeO_4$	25	0.54	−0.39	0.055	0.28−6.79
	2	Na_2SeO_4	25	0.45	−0.25	0.04	0.66−2.15
$CdBr_2$	4	$BaBr_2$	25	0.015	0.0175	0	0.3−2.55
	0		25	−0.022	0.012	0	2.55−3.67
	4		30	0.015	0.013	0	0.77−1.94
	0		30	−0.022	0.012	0	1.94−4.14
	4		35	0	0	0	0.2792
	0		35	−0.022	0.012	0	0.62−4.25
	0		50	−0.022	0.011	0	0.86−4.74
	0		75	−0.022	0.011	0	0.86−5.61
	0		100	−0.022	0.011	0	0.59−6.39
$CdCl_2$	1	NaCl	100	0.013	0.0013	0	1.12−6.59
	1	LiCl	40	−0.0055	0	0	2.5−8.69
$Cd(NO_3)_2$	9	HNO_3	0	0.017	0	0	0.03−2.24
	4		0	−0.008	0.0003	0	2.24−51.01
	4		20	0.004	0	0	2.78−20.47
	4		25	−0.001	0.0003	0	0.06−34.2
	4		15	−0.007	0.0003	0	3.72−43.9
$CdSO_4$	$\frac{8}{3}$	$MgSO_4$	25	−0.038	0	0	0.62−2.12
	$\frac{8}{3}$		40	0.050	0	0	0.70−1.43
	1	$(NH_4)_2SO_4$	50	0.073	0	0	0.74−1.22
	1		77.1	0.105	−0.001	0	0.45−1.47
	1		97	0.105	0	0	0.54−1.12

contd.

Component 1	H$_2$O	Component 2	t °C	Q_1	Q_2	Q_3	Data for c_2
CdSO$_4$	0	Na$_2$SO$_4$	50	0.020	0	0	0.41—1.01
	0		65	0.034	0	0	0.40—1.26
	0		80	0.050	0	0	0.22—1.20
	0		97	0.102	−0.017	0	0.27—1.19
CoCl$_2$	6	CaCl$_2$	0	−0.035	0.0015	0	3.58—5.48
	6		25	0.006	−0.0045	0	0.78—2.9
	6		25	−0.030	0.005	0	1.27—6.07
	2		25	−0.021	0	0	6.07—7.46
	6		50	0.017	0	0	0.5—1.09
	2		50	−0.015	0.0015	0	1.09—10.35
	6	CdCl$_2$	25	0.074	−0.004	0	0.48—3.49
	6	CoSO$_4$	25	−0.031	0	0	0.37—0.47
	6		38	−0.052	−0.029	0	0.4—0.52
	6	HCl	25	−0.033	0.0022	0	2.19—10.69
	2		25	−0.033	0.0018	0	10.69—17.8
	6		0	−0.050	0.0015	0	1.2—13.12
	6	HgCl$_2$	25	0.028	0.0012	0	1.02—6.84
	6	KCl	25	0.0375	0	0	0.52—1.91
	6	LiCl	0	−0.039	0.0024	0	7.0—12.8
	6		25	−0.022	0.0020	0	2.8—9.9
	2		25	−0.077	0.0069	0	5.6—8.56
	6		40	−0.018	0.0039	0	1.32—6.0
	2		40	−0.030	0.0021	0	6.0—15.77
	2		45	−0.020	0.0013	0	5.5—14.67
	2		80	−0.013	0.0012	0	8.17—17.32
	6	MgCl$_2$	25	−0.046	0.0020	0	1.0—3.7
	6	NaCl	20	0.014	0	0	0.9—0.98
	6		25	0.010	0	0	0.48—1.15
	2		60	0.030	0	0	0.09—0.82
	2		100	0.016	0	0	0.8—2.22
	6	RbCl	25	0.057	0	0	0.45—0.84
	6	ZnCl$_2$	25	0.030	0.005	0	1.15—5.0
	2		25	0.013	0.0009	0	5.0—6.56
Co(NO$_3$)$_2$	6	Cu(NO$_3$)$_2$	14	0.012	−0.0012	0	0.5—4.21
	6		20	−0.001	0.0015	0	1.07—6.26
	6		30	−0.008	0.0010	0	1.0—5.06
	6	HNO$_3$	25	−0.001	0.0010	0	0.4—11.59
	3		80	0.036	−0.0030	0	1.3—6.25
CoSO$_4$	7	CoCl$_2$	0	−0.078	0.01	0	0.4—4.6
	7		25	−0.078	0.011	0	0.11—5.49
	7		38	−0.04	0	0	0.6—2.11
	6		38	−0.04	0.0018	0	2.11—6.47

contd.

Component 1	H$_2$O	Component 2	t °C	Q_1	Q_2	Q_3	Data for c_2
CoSO$_4$	6	CoCl$_2$	50	−0.04	0.0035	0	0.96−5.97
	1		75	−0.045	0.0018	0	0.22−8.3
	1		99.5	−0.052	0.0020	0	0.42−11.75
	7	H$_2$SO$_4$	0	−0.003	0.0015	0	0.35−7.93
	7		20	−0.003	0.0025	0	0.33−4.40
	7		25	0.01	0.0015	0	0.73−3.64
	7		40	0.01	0.0016	0	0.34−1.66
	6		40	0.004	0.0024	0	1.07−5.9
	1		70	0.004	−0.0055	0	0.80−11.09
	1		80	0.018	−0.004	0	3.37−8.84
	7	Na$_2$SO$_4$	25	0.130	−0.08	0	0.07−1.22
	1		97	−0.062	0	0	0.31−1.02
CsNO$_3$	0	Pb(NO$_3$)$_2$	25	0.205	−0.026	0	0.17−2.82
Cs$_2$SO$_4$	0	Ag$_2$SO$_4$	25	0.052	−0.02	0	0.05−0.14
CsCl	0	LiCl	25	0.002	0.0007	0	2.3−11.6
	0		40	−0.005	0.0007	0	4.4−11.5
CuBr$_2$	0	HBr	25	−0.0166	0.0008	0	0.79−6.0
CuCl	0	FeCl$_2$	25	1.00	−0.41	0	0.47−2.31
CuCl$_2$	2	CdCl$_2$	25	0.031	0	0	0.85−2.67
	2	CaSO$_4$	30	0.01	0	0	0.12−0.33
	2	HCl	0	−0.03	0.002	0	1.65−6.77
	2		25	−0.025	0.002	0	2.94−7.79
	2		75	0.046	−0.002	0	1.7−4.36
	2	HgCl$_2$	25	0.026	0.0012	0	6.25−7.08
	2		35	0.022	0.0015	0	1.39−6.6
	2	LiCl	0	−0.019	0	0	2.2−11.76
	2		19	−0.024	0.0015	0	2.5−11.54
	2		25	−0.024	0	0	1.8−8.55
	2		30	−0.025	0.002	0	1.02−11.4
	2		50	−0.015	0.0012	0	2.8−11.35
	2		99	−0.007	0.001	0	2.5−10.88
	2	NaCl	30	0.009	0	0	0.95−3.31
Cu(NO$_3$)$_2$	6	Co(NO$_3$)$_2$	14	−0.006	0	0	0.68−1.4
	6		20	−0.006	0.028	0	1.17−1.48
	6		30	−0.071	0.021	0	0.72−2.38
	6	CuSO$_4$	20	0.11	0	0	0.15−0.17
	2.5	NH$_4$NO$_3$	30	0.026	−0.0004	0	5.06−30.48
	6	NaNO$_3$	20	0.067	−0.0289	0	0.57−1.49

contd.

Component 1	H_2O	Component 2	t °C	Q_1	Q_2	Q_3	Data for c_2
$CuSO_4$	5	$BeSO_4$	30	0.032	−0.012	0	0.12−3.66
	5		80	0.048	−0.012	0	0.33−5.34
	5	$CoSO_4$	0	0.247	−0.14	0	0.07−0.46
	5		25	0.130	−0.08	0	0.09−1.22
	5	$CuCl_2$	30	−0.098	0.0155	0	0.61−5.81
	5	$Cu(NO_3)_2$	20	−0.098	0.012	0	0.76−6.69
	5		35	−0.122	0.012	0	0.78−8.44
	5	$FeSO_4$	30	0.008	0.045	0	0.2−1.12
	5	Li_2SO_4	0	0.064	−0.007	0	0.53−3.06
	5		25	0.052	−0.007	0	0.87−2.79
	5		30	0.029	−0.0035	0	0.4−2.69
	5		55	0.013	0	0	0.4−2.53
	5	$MgSO_4$	0	0.169	−0.055	0	0.71−1.93
	5	Na_2SO_4	20	0.130	−0.008	0	0.55−1.14
	5		97	0	0	0	0.11−0.39
$CuSeO_4$	5	H_2SeO_4	25	0.12	−0.03	0	0.5−3.5
	5	$(NH_4)_2SeO_4$	25	0.25	0	−0.03	0.34−0.7
$FeCl_2$	4	$CuCl$	25	0.045	0	0	1.2−2.34
	4	HCl	0	−0.037	0.0008	0	6.49−12.87
	4		20	−0.0255	0	0	0.79−12.6
	4		40	−0.027	0.0010	0	2.3−11.38
	4		60	−0.012	0	0	2.45−6.84
	2		60	−0.0218	0	0	6.84−11.8
	2		100	−0.0208	0	0	0.68−6.08
$FeSO_4$	7	$BeSO_4$	25	−0.005	0	0	0.71−3.36
	4		60	−0.004	−0.015	0	1.27−3.04
	7	H_2SO_4	−10	0.180	−0.065	0.006	0.0−3.5
	7		−5	0.156	−0.065	0.008	0.7−3.5
	7		0	0.128	−0.065	0.011	0.0−3.0
	7		0	0.023	0	0	3.0−7.8
	7		5	0.098	−0.053	0.011	0.16−2.2
	7		5	0.023	0	0	2.2−4.0
	7		10	0.084	−0.038	0.008	0.0−2.2
	7		10	0.025	0	0	2.2−4.0
	7		15	0.057	−0.026	0.007	0.0−2.2
	7		15	0.025	0	0	2.2−4.0
	7		20	0.047	−0.017	0.005	0.0−2.2
	7		20	0.023	0	0	2.2−4.0
	7		25	0.025	0	0	0.0−5.1
	7		30	0.025	0	0	0.8−3.6
	7		35	0.009	0.034	−0.01	0.0−1.0
	7		35	0.025	0	0	1.0−3.4
	7		40	0	0.034	−0.01	0.0−1.5

contd.

Component 1	H$_2$O	Component 2	t °C	Q_1	Q_2	Q_3	Data for c_2
FeSO$_4$	7	H$_2$SO$_4$	40	0.025	0	0	1.5−5.0
	1		0	−0.115	0.0023	0	2.0−40
	1		25	−0.09	0.0027	0	2.0−20
	1		45	−0.08	0.0023	0	2.0−5.0
	1		55	−0.048	0.0011	0	2.0−24
	1		60	−0.034	−0.0008	0	2.0−16
	1		100	0.055	−0.0035	0	1.5−11
	1		100	−0.013	0.056	−0.014	11−40
	7	Li$_2$SO$_4$	30	0.024	0	0	0.5−2.22
	1	Na$_2$SO$_4$	97	0	−0.008	0	0.16−1.64
KBr	0	HBr	25	−0.057	0	0	1.3−3.47
	0	HgBr$_2$	0	0.190	−0.017	0	1.16−4.00
	0		34	0.205	−0.05	0.003	0.0−100
	0		80	0.150	−0.05	0.004	0−75
	0	KBrO$_3$	25	−0.0255	0.036	0	0.12−0.14
	0		40	0.005	0.041	0	0.03−0.23
	0		80	−0.013	0.039	0	0.7
	0	K$_2$Cr$_2$O$_7$	0	−0.0252	0	0	0.0156
	0		10	0.0751	0	0	0.0276
	0		20	0.0811	0	0	0.0367
	0		30	0.037	0	0	0.0556
	0		40	0.0546	0	0	0.1003
	0	KIO$_3$	5	0.065	−0.024	0	0.1345
	0		25	0.052	−0.024	0	0.07−0.19
	0		50	0.046	−0.024	0	0.19−0.27
	0	KNO$_3$	25	0.064	−0.025	0	0.5−1.85
	0	KOH	17	−0.032	0	0	0.6−14.4
	0	K$_2$SO$_4$	20	−0.114	0	0	0.02−0.05
	0	NaBr	−10	−0.02	−0.0015	0	1.0−7.0
	0		0	−0.02	−0.0010	0	1.0−7.5
	0		10	−0.02	−0.0012	0	1.0−8.0
	0		20	−0.027	0	0	1.0−8.5
	0		25	−0.027	0	0	1.0−8.8
	0		30	−0.012	−0.0012	0	1.0−9.3
	0		50	−0.019	0	0	2.4−7.48
	0	PbBr$_2$	25	0.135	0	0	0.08−0.36
KBrO$_3$	0	KBr	0	0.125	−0.005	0	4.6
	0		20	0.0825	−0.005	0	1.15−5.54
	0		25	0.075	−0.005	0	0.7−5.7
	0		40	0.036	−0.0011	0	0.8−6.4
	0		60	0.032	−0.0025	0	0.4−7.42
	0		80	0.017	−0.0009	0	0.87−8.16

contd.

Component 1	H$_2$O	Component 2	t °C	Q_1	Q_2	Q_3	Data for c_2
KBrO$_3$	0	KCl	25	0.068	-0.0055	0	0.6 — 4.8
	0	KI	25	0.070	-0.0032	0	0.6 — 8.95
	0	KNO$_3$	25	0.148	-0.012	0	0.55 — 3.85
	0	K$_2$SO$_4$	25	0.2865	-0.153	0	0.16 — 0.62
KCN	0	TlCN	25	-0.155	-0.43	0	0.05 — 0.23
	0	AgCN	25	0.052	0	0	0.92 — 3.74
K$_2$CO$_3$	0	Li$_2$CO$_3$	25	0.057	-0.17	0	0.1 — 0.16
	6	KOH	-20	-0.01	0	0	2.36 — 5.67
	6		-10	-0.01	0	0	1.35 — 2.4
	1.5		-10	-0.02	-0.0006	0	1.78 — 15.49
	1.5		0	-0.02	-0.0005	0	2.4 — 16.79
	1.5		10	-0.02	-0.0003	0	3.4 — 17.82
	1.5		25	-0.02	0	0	3.21 — 20.7
	1.5		30	-0.0155	0	0	21.9 — 22.95
K$_2$C$_2$O$_4$	1	Na$_2$C$_2$O$_4$	25	-0.033	0.2	0	0.07 — 0.26
KCl	0	BeCl$_2$	25	-0.090	0.004	0	0.2 — 6.6
	0	CaCl$_2$	20	-0.102	0.0045	0	0.75 — 7.10
	0		25	-0.071	0.0050	0	1.04 — 7.69
	0		30	-0.071	0.0070	0	0.28 — 9.55
	0		35	-0.056	0.0060	0	0.38 — 10.56
	0		50	-0.051	0.0055	0	0.76 — 11.29
	0		95	-0.046	0.0050	0	0.89 — 11.76
	0	HgCl$_2$	34	0.170	-0.025	0	0.24 — 1.42
	0	CoCl$_2$	0	-0.012	0	0	0.53 — 4.5
	0		20	-0.008	0	0	1.05 — 5.78
	0		25	0	0	0	1.08 — 5.94
	0		38	0.008	0	0	1.15 — 8.15
	0		50	-0.002	0.0021	0	0.56 — 9.23
	0		75	-0.0002	0.0034	0	0.45 — 11.8
	0		99.5	0.036	0	0	1.28 — 18.1
	0	CuCl$_2$	0	-0.002	0.0175	0	0.6 — 2.27
	0		25	0.0455	0	0	0.57 — 2.65
	0		75	0.0585	0	0	1.63 — 5.85
	0	HCl	0	-0.068	0.0032	0	0.7 — 12.3
	0		25	-0.069	0.0030	0	6.18 — 17.22
	0	KBr	10	-0.016	-0.0014	0	0.9 — 3.24
	0		15	-0.0135	-0.007	0	0.9 — 3.26
	0		20	-0.011	0.0004	0	0.9 — 3.26
	0	KBrO$_3$	25	0.0367	0	0	0.12 — 0.13

contd.

Component 1	H$_2$O	Component 2	t °C	Q_1	Q_2	Q_3	Data for c_2
KCl	0	K$_2$CO$_3$	30	−0.047	0	0	7.9−8.1
	0	KClO$_3$	30	0.028	0	0	0.08−0.26
	0		50	0.026	0	0	0.2−0.52
	0		70	0.0185	0	0	0.3−0.95
	0		100	0.013	0	0	1.0−2.11
	0		150	0.009	0	0	1.5−5.41
	0		175	0.0085	0	0	2.0−10.6
	0		200	0.007	0	0	2.04−11.66
	0	KClO$_4$	150	0.014	0	0	0.43−1.57
	0		175	0.014	−0.0022	0	0.7−2.85
	0		200	0.014	−0.002	0	0.7−3.8
	0		225	0.014	−0.001	0	2.4−7.75
	0	K$_2$Cr$_2$O$_7$	25	0.111	0	0	0.04−0.05
	0		50	0.023	0.012	0	0.09−0.21
	0		100	−0.032	0.032	0	0.35−0.98
	0	KF	25	−0.019	0	0	0.9−18.3
	0		75	−0.016	0	0	1.2−19.13
	0	KHCO$_3$	10	0.0075	0	0	0.5−1.25
	0		20	0.0042	0	0	0.5−1.52
	0		29.85	0.0026	0	0	1.8222
	0		40	0.0012	0	0	2.1978
	0	KHF$_2$	25	−0.002	0.003	0	0.7−3.59
	0		75	−0.002	0.001	0	0.861−11.5
	0	KI	0	−0.005	0	0	0.19−7.09
	0		25	−0.030	0.006	0	0·2−3.7
	0		50	−0.020	0.001	0	0.4−9.2
	0		75	−0.020	0.0007	0	0.3−9.5
	0	KIO$_3$	5	0.0893	0	0	0.0886
	0		25	0.0466	0	0	0.1361
	0		50	0.0280	0	0	0.121
	0	KNO$_3$	0	0.064	−0.006	0	0.25−0.94
	0		17.5	0.034	−0.003	0	0.75−1.84
	0		18.5	0.034	−0.002	0	0.2−1.8
	0		25	0.031	0	0	0.6−2.38
	0		30	0.029	0	0	0.8−2.6
	0		40	0.023	0	0	1.6−3.9
	0		91	0.017	−0.0004	0	5.2−16.12
	0		150	0.013	−0.0002	0	0.97−21.61
	0		200	0.013	−0.0002	0	1.33−15.76
	0		250	0.013	−0.0002	0	0.09−0.88
	0	K$_2$SO$_4$	0	0.1	0	0	0.03−0.06
	0		15	0.06	0	0	0.06−0.08
	0		20	0.063	0	0	0.03−0.09
	0		30	−0.078	0	0	0.02−0.09
	0		50	−0.088	−0,0067	0	0,1042

contd.

Component 1	H₂O	Component 2	t °C	Q_1	Q_2	Q_3	Data for c_2
KCl	0	LiCl	25	−0.030	−0.001	0	1.29—20.75
	0		30	−0.023	0.0009	0	1.8—18.5
	0		40	−0.032	0.0013	0	1.8—17.1
	0	MgCl₂	0	−0.072	−0.0045	0	0.7—4.0
	0		25	−0.257	−0.011	0	1.1—4.0
	0		35	−0.061	−0.0036	0	0.5—4.2
	0		55	−0.024	−0.005	0	1.0—4.5
	0		75	−0.053	−0.0026	0	1.0—4.7
	0		83	−0.031	−0.0063	0	1.0—4.8
	0		100	−0.043	−0.0044	0	0.7—4.8
	0		105	−0.017	−0.0065	0	1.0—5.2
	0	NH₄Cl	0	−0.010	0	0	1.9—4.49
	0		20	−0.010	0	0	3.77—5.96
	0		25	−0.011	0	0	2.2—6.11
	0		30	−0.012	0	0	1.07—6.5
	0		45	−0.005	0	0	1.4—7.41
	0		65	−0.005	0	0	2.9—9.06
	0		90	−0.003	0	0	3.8—11.08
	0	NaCl	0	−0.015	0	0	1.7—5.45
	0		20	−0.015	0	0	1.7—4.91
	0		25	−0.015	0	0	2.96—5.1
	0		40	−0.015	0	0	1.7—5.0
	0		60	−0.0135	0	0	1.7—4.8
	0		80	−0.012	0	0	1.7—4.75
	0		100	−0.0105	0	0	1.7—4.4
	0		120	−0.0075	0	0	1.7—4.6
	0		140	−0.006	0	0	1.7—4.7
	0		189.5	−0.003	0	0	1.7—5.4
	0		200	0	0	0	1.7—5.45
	0	SrCl₂	18	−0.046	0.0032	0	1.2—3.16
	0		60	−0.038	0.004	0	1.8—5.6
	0		100	−0.026	0.0039	0	1.8—6.9
KClO₃	0	Ca(ClO₃)₂	20	0.155	−0.009	0	0.62—8.98
	0	KBr	25	0.0455	−0.0025	0	0.9—5.75
	0	K₂CO₃	24.2	0.153	0	0	0.7—8.1
	0		40	0.0365	−0.002	0	1.76—7.94
	0	KCl	0	0.134	−0.013	0	1.25—3.63
	0		10	0.0755	−0.003	0	0.079—4.19
	0		20	0.053	−0.003	0	0.4—4.47
	0		30	0.036	−0.0015	0	0.1—5.0
	0		40	0.030	−0.0015	0	0.7—3.2
	0		50	0.016	0	0	1.4—5.4
	0		70	−0.005	0	0.0012	0.4—6.25
	0		150	−0.004	−0.0003	0	1.5—5.56
	0		175	0.003	0	0	3.5—7.95
	0		200	0.003	0	0	1.6—2.8

contd.

Component 1	H$_2$O	Component 2	t °C	Q_1	Q_2	Q_3	Data for c_2
KClO$_3$	0	KI	25	0.0375	−0.001	0	0.6−8.95
	0	KIO$_3$	25	0.154	−0.07	0	0.15−0.29
	0		50	0.068	−0.018	0	0.13−0.43
	0	K$_2$SO$_4$	15	0.262	−0.11	0	0.13−0.5
	0		25	0.224	−0.11	0	0.17−0.57
	0		45	0.155	−0.11	0	0.6464
	0	NaClO$_3$	10	0.108	−0.006	0	0.07−8.27
	0		30	0.053	0	0	0.16−9.97
	0		50	0.027	0	0	0.4−11.96
	0		70	0.015	0.0004	0	0.87−14.45
	0		100	0.010	0.0002	0	4.46−19.9
K$_2$CrO$_4$	0	K$_2$MoO$_4$	25	−0.027	0	0	0.64−7.68
K$_2$Cr$_2$O$_7$	0	CaCr$_2$O$_7$	50	0.096	−0.0049	0	0.6−7.22
	0	KBr	0	0.160	−0.014	0	0.6−4.4
	0		10	0.045	0.005	0	0.6−5.16
	0		20	0.012	0.005	0	0.37−5.46
	0		30	−0.003	0.005	0	0.6−5.91
	0		40	−0.022	0.005	0	0.7−6.37
K$_2$Cr$_2$O$_7$	0	KCl	25	0.018	0.003	0	0.32−4.78
	0		50	−0.015	0.0025	0	0.44−5.78
	0		100	−0.024	0.0025	0	0.86−6.58
	0	Na$_2$Cr$_2$O$_7$	25	0.305	−0.050	0	0.2−3.21
	0		50	0.100	−0.005	0	0.35−5.25
	0		100	0.010	0.0025	0	0.46−8.57
KHCO$_3$	0	K$_2$CO$_3$	5	0	0	0	6.5−7.4
	0		25	0	−0.0007	0	1.02−7.59
	0		35	0	−0.0009	0	0.1−7.75
	0		50	0.023	−0.0035	0	0.5−8.1
	0	KCl	10	−0.016	0.0008	0	1.2−3.68
	0		20	−0.019	0.0009	0	1.34−3.98
	0		29.85	−0.017	0	0	4.1973
	0		40	−0.017	0	0	4.4293
	0	NaHCO$_3$	20	0.0285	0	0	0.2−0.7
	0		25	0.0250	0	0	0.18−0.77
	0		30	0.022	0	0	0.3−0.8
	0		50	0.042	0	0	0.8−1.5
KHF$_2$	0	KCl	25	−0.022	0.004	0	0.8−3.59
	0		75	−0.018	0.0043	0	0.95−4.19
KH$_2$PO$_4$	0	NH$_4$NO$_3$	0	0.184	−0.0356	0	0.51
	0		10	0.142	−0.026	0	0.5−1.35
	0		20	0.091	−0.015	0	0.5−2.0
	0		30	0.057	−0.0032	0	0.5−2.4

contd.

Component 1	H_2O	Component 2	t °C	Q_1	Q_2	Q_3	Data for c_2
K_2HPO_4	0	Na_2HPO_4	25	0.02	—0.0025	0	0.44—1.43
KI	0	AgI	30	0.065	—0.0023	0	2.39—14.79
	0		50	0.065	—0.0023	0	1,5—25.27
	0	HgI_2	20	0.091	—0.004	0	1.4—8.12
	0		22.5	0.066	—0.0017	0	0.08—9.19
	0	$KBrO_3$	25	0.0272	0	0	0.1443
	0	KCl	0	—0.023	0	0	1.239
	0		25	—0.0065	0	0	1.48—1.51
	0		50	—0.0028	0	0	1.72—1.79
	0		75	—0.007	0	0	1.99—2.06
	0	$KClO_3$	25	0.026	0	0	0.13—0.17
	0	KIO_3	25	0.028	0.023	0	0.12—0.28
	0	NaI	8	—0.025	—0.0005	0	3.13—10.75
	0		25	—0.025	0	0	1.53—11.97
	0		40	—0.025	0	0	2.33—13.66
KIO_3	0	KBr	5	0.15	—0.009	0	0.7—4.8
	0		25	0.091	—0.009	0	0.8—5.77
	0		50	0.044	0	0	0.8—6.85
	0	KCl	5	0.131	—0.0094	0	0.4—4.0
	0		25	0.071	—0.002	0	0.4—4.8
	0		50	0.032	0.001	0	1.0—5.75
	0	$KClO_3$	25	0.119	0.005	0	0.2—0.63
	0		50	0.066	0.003	0	0.35—1.42
	0	KI	25	0·095	—0.0033	0	0.15—8.44
	0	K_2MoO_4	25	0.192	—0.0305	0.0019	0.1—0.49
	0		25	0.243	—0.125	0.058	0.49—7.7
	0	KNO_3	5	0.248	—0.0324	0	0.58—1.68
	0		25	0.157	—0.0132	0	0.6—3.85
	0		50.4	0.085	—0.004	0	2.29—8.57
	0	K_2SO_4	5	0.379	0	0	0.25—0.45
	0		25	0.273	—0.093	0	0.3—0.64
	0		50	0.206	—0.08	0	0.54—0.85
	0	$NaIO_3$	5	0.41	0	0	0.07—0.11
	0		25	0.205	0	0	0.24—0.41
	0		50	0.15	0	0	0.23—0.7
KIO_4	0	KF_4B	35	0.73	0	0	0.01—0.07
K_2MoO_4	0	KIO_3	25	0.0346	0	0	0.1567
	0	KNO_3	25	0.0285	—0.0005	0	0.91—1.59

contd.

Component 1	H$_2$O	Component 2	t °C	Q_1	Q_2	Q_3	Data for c_2
K$_2$MoO$_4$	0	KOH	25	−0.013	−0.0001	0	1.1−11.82
KNO$_2$	0	TlNO$_2$	25	0.014	−0.0002	0	1.02−12.49
KNO$_3$	0	AgNO$_3$	30	0.08	−0.0035	0	1.16−9.54
	0	Ca(NO$_3$)$_2$	0	0.164	−0.0078	0	1.74−7.3
	0		20	0.076	−0.0008	0	0.96−8.65
	0		25	0.058	0	0	0.17−7.0
	0		30	0.058	0	0	0.8−7.0
	0	KBr	25	0.010	0	0	0.95−5.73
	0	KBrO$_3$	25	0.093	−0.041	0	0.2−0.3
	0	K$_2$CO$_3$	24.2	−0.030	0	0	0.94−8.05
	0	KCl	10	0.0175	0.002	0	2.1−4.1
	0		18.5	−0.006	0.004	0	0.2−4.4
	0		20.5	−0.006	0.0045	0	0.7−4.1
	0		25	−0.014	0.0043	0	0.8−4.65
	0		30	−0.02	0.0043	0	1.1−4.7
	0		40	−0.024	0.0038	0	1.5−5.05
	0		91	−0.015	0.0016	0	2.0−5.3
	0		150	−0.008	0.002	0	0.9−7.44
	0	KClO$_3$	25	0.177	−0.168	0	0.18−0.46
	0	K$_2$CrO$_4$	0	0.132	−0.018	0	0.6−2.94
	0		25	0.002	0	0	0.61−2.89
	0		50	−0.03	0.0045	0	0.61−2.83
	0	KH$_2$PO$_4$	0	0.061	0	0	0.3−0.88
	0		10	0.023	0	0	0.3−0.80
	0		20	0	0	0	0.29−1.06
	0		30	−0.035	0	0	0.3−1.21
	0	KIO$_3$	5	0.12	0	0	0.05−0.16
	0		25	0.075	0	0	0.13−0.17
	0		50.4	0.023	0	0	0.21−0.43
	0	K$_2$MoO$_4$	25	−0.034	0.0065	0	0.9−7.97
	0	K$_2$SO$_4$	0	0.159	0	0	0.2−0.28
	0		25	−0.023	0.35	0	0.13−0.32
	0		35	−0.10	0	0	0.08−0.29
	0	KVO$_3$	25	0.013	0	0	0.02−0.08
	0	MgCl$_2$	−10	0.108	−0.0285	0	2.0−3.75
	0		0	−0.081	−0.0012	0	1.14−3.8
	0		10	−0.116	0	0	1.16−3.81
	0		20	−0.191	0.021	0	1.13−3.82
	0		30	−0.198	0.021	0	1.14−3.85

contd.

Component 1	H_2O	Component 2	t °C	Q_1	Q_2	Q_3	Data for c_2
KNO_3	0	$Mg(NO_3)_2$	0	−0.006	0.028	0	0.17−2.69
	0		25	−0.086	0.024	0	0.5−2.76
	0		50	−0.108	0.017	0	0.53−6.12
	0		99.5	−0.057	0.0045	0	2.4−8.46
	0	NaCl	20	0.016	−0.0007	0	1.7−6.45
	0		30	0.004	0	0	1.68−6.62
	0		40	0	0	0	1.9−6.63
	0		91	0.0035	0	0	2.1−6.68
	0	$NaNO_3$	0	0.078	−0.0018	0	9.064
	0		20	0.056	−0.0018	0	12.1027
	0		25	0.047	−0.0014	0	11.6731
	0		30	0.038	−0.0009	0	3.04−12.15
	0		40	0.025	−0.0003	0	2.8−13.63
	0		50	0.015	0	0	4.99−14.95
	0		75	0.01	0	0	6.5−19.9
	0		87.5	0.01	0	0	23.3323
	0		91	0.0088	0	0	5.1−24.1
	0		100	0.0085	0	0	9.3−28.39
	0	$Pb(NO_3)_2$	0	0.373	−0.057	0	1.09−2.33
	0		25	0.216	−0.030	0	0.79−3.79
	0		50	0.090	−0.005	0	0.78−5.14
	0		100	0.011	0	0	3.7597
	0	$Sr(NO_3)_2$	20	0.126	−0.0082	0	0.36−4.58
	0		40	0.0525	0	0	2.4−5.25
	0	$UO_2(NO_3)_2$	0	0.012	0.027	0	0.02−2.5
	0		5	0.003	0.022	0	0.02−2.78
	0		10	−0.033	0.030	0	0.02−3.7
	0		15	−0.033	0.028	0	0.02−3.39
	0		20	−0.10	0.063	−0.006	0.02−3.6
	0		25	−0.146	0.060	0	0.02−3.01
	0	$CO(NH_2)_2 \cdot HNO_3$	25	0	0	0	0.2−1.4
	0		40	−0.005	0	0	0.4−2.2
	0		60	−0.108	0.05	−0.0044	0.1−3.2
KOH	4	K_2CO_3	−60	−0.060	0	0	0.13−1.1
	4		−40	0	0.009	0.002	0.2−1.4
	2		−10	0	0.020	0	0.12−0.26
	2		0	0	−0.057	0	0.3026
	2		10	0	−0.0168	0	0.3266
KSCN	0	AgSCN	25	0.035	0	0	1.18−2.31
	0	$Ba(SCN)_2$	25	0	0	0	1.47−4.9
	0	KNO_3	25	0.0135	0	0	2.0−3.7
	0	K_2SO_4	25	−0.1390	0	0	0.0078
	0		40	−0.1429	0	0	0.0067

contd.

Component 1	H$_2$O	Component 2	t °C	Q_1	Q_2	Q_3	Data for c_2
K$_2$SO$_4$	0	Ag$_2$SO$_4$	25	0.430	0	0	0.02 — 0.04
	0	BeSO$_4$	0	0.335	0.015	0	0.14 — 0.73
	0		50	0.229	—0.06	0	0.21 — 0.64
	0		75	0.229	—0.07	0	0.14 — 0.71
	0		99.5	0.229	—0.05	0	0.45 — 0.86
	0	CdSO$_4$	0	0.33	—0.035	0	0.97 — 1.40
	0		25	0.281	—0.055	0	0.14 — 1.81
	0		50	0.224	—0.035	0	0.18 — 1.03
	0		75	0.202	—0.035	0	0.42 — 0.79
	0		90	0.224	—0.035	0	0.38 — 0.71
	0		99	0.224	—0.035	0	0.33 — 0.64
	0	KBr	20	0.016	—0.0012	0	0.42 — 5.43
	0	KBrO$_3$	25	0.025	—0.040	0	0.11 — 0.27
	0	K$_2$CO$_3$	25	0.026	—0.005	0	0.45 — 8.1
	0		50	0.026	—0.003	0	0.98 — 8.54
	0		150	0.026	—0.002	0	1.87 — 15.91
	0	KCl	0	0.061	—0.004	0	0.48 — 3.73
	0		15	0.045	—0.0045	0	0.58 — 4.34
	0		20	0.045	—0.0045	0	0.65 — 4.56
	0		25	0.020	—0.0045	0	1.23 — 4.78
	0		30	0.020	—0.0020	0	0.43 — 4.84
	0		40	0	0	0	0.44 — 5.52
	0		70	0	0	0	0.62 — 6.23
	0	KClO$_3$	15	0.054	—0.030	0	0.3022
	0		25	0.054	—0.030	0	0.16 — 0.47
	0		45	0.054	—0.030	0	0.9864
	0	KHSO$_4$	50	0.032	—0.003	0	0.28 — 5.2
	0	KI	25	—0.073	0.035	—0.003	0.65 — 8.89
	0	KIO$_3$	5	0.125	0	0	0.09 — 0.13
	0		25	0.068	0	0	0.13 — 0.23
	0		50	0.0374	0	0	0.21 — 0.40
	0	KNO$_3$	0	0.14	—0.0065	0	0.63 — 1.20
	0		25	0.076	—0.0040	0	0.55 — 3.55
	0		35	0.06	—0.0033	0	2.08 — 4.61
	0		50	0.04	—0.0012	0	1.78 — 8.15
	0		75	0.0286	—0.0007	0	1.76 — 12.97
	0		100	0.018	—0.0002	0	2.5 — 23.47
	0	KSCN	25	—0.083	0.054	—0.005	0.13 — 1.88
	0		40	—0.12	0.054	—0.006	0.05 — 6.05
	0	MgSO$_4$	25	0.230	—0.056	0	0.24 — 1.37
	0		30	0.230	—0.059	0	0.39 — 1.46
	0		35	0.230	—0.070	0	0.37 — 1.53

contd.

Component 1	H_2O	Component 2	t °C	Q_1	Q_2	Q_3	Data for c_2
K_2SO_4	0	$MgSO_4$	50	0.151	−0.022	0	0.5−2.01
	0		66	0.11	−0.014	0	0.83−2.07
	0		55	0	9	0	0.19−2.19
	0		75	0.11	−0.016	0.005	0.93−2.16
	0		100	0.11	0	0	0.83−1.79
	0	$MnSO_4$	0	0.45	−0.14	0	0.47−1.16
	0		25	0.334	−0.14	0	0.51−1.54
	0		50	0.334	−0.14	0	0.22−1.07
	0		80	0.248	−0.14	0	0.28−0.51
	0		97	0.24	−0.15	0	0.17−0.31
	0	Na_2SO_4	35	0.275	−0.18	0	0.07−0.49
	0	$ZnSO_4$	25	0.36	0	0	0.07−0.13
	0		80	0.248	−0.086	0	0.16−1.24
K_2SeO_4	0	$CaSeO_4$	25	−1.513	0	0	0.1096
	0	Na_2SeO_4	25	−0.108	−0.01	0	0.8−3.53
$K_2S_2O_6$	0	BaS_2O_6	0	0.80	0	0	0.17−0.33
	0		20	0.45	−0.165	0	0.48−0.75
	0		30	0.395	−0.145	0	0.23−0.97
KVO_3	0	KCl	25	−0.20	0.027	0	2.04−4.82
	0	KNO_3	25	−0.195	0.0752	0	0.7−2.1
	0		25	−0.060	0.0093	0	2.1−4.97
	0	K_2SO_4	25	−0.174	0	0	0.07−0.74
LiBr	2	HBr	25	−0.008	0	0	1.71−10.36
	2	$PbBr_2$	25	0.015	−0.001	0	1.2−3.87
LiCl	1	$BaCl_2$	25	−0.058	0	0	1.30−10.38
	1		40	−0.0025	0	0	1.10−6.27
	1	$BeCl_2$	0	−0.025	−0.005	0	1.0−6.0
	1		25	−0.025	−0.003	0	1.24−5.9
	1		40	−0.025	−0.003	0	1.5−9.77
	1	$CdCl_2$	25	−0.006	0.014	0	0.20−1.49
	1		40	−0.006	0.006	0	0.46−2.29
	1	CsCl	25	−0.0075	0.0030	0	0.6−7.96
	1		40	0.0040	0.0028	0	0.4−7.5
	2	$CuCl_2$	0	−0.0643	0	0	0.25
	1		25	0.024	−0.028	0	0.2−0.8
	1		30	0.018	0	0	0.23−0.25
	1		50	0.012	0	0	0.15−0.45
	0		99	0.0034	0	0	1.37−1.66
	0	HCl	0	0	0	0	4.46−17.62
	0		25	0.0022	0	0	1.54−4.23

contd.

Component 1	H_2O	Component 2	t °C	Q_1	Q_2	Q_3	Data for c_2
LiCl	1	$HgCl_2$	30	−0.154	0.089	−0.016	0.17−3.57
	1	KCl	25	0.094	−0.15	0	0.57−0.83
	1		30	0.094	−0.034	0	0.2−0.5
	1		40	0.094	−0.190	0	0.4−0.63
	2	$LiClO_3$	3	0.013	0	0	5.6−11.9
	1		3	0.007	0	0	11.9−22.08
	2		8.5	0.014	0	0	1.13−8.13
	1		8.5	0.013	−0.0003	0	8.13−26.98
	2		6	0.018	−0.0004	0	3.59−10.21
	1		6	0.010	−0.0002	0	10.21−24.4
	1		25	0.012	−0.0002	0	6.5−35
	3	$MgCl_2$	−50	0.025	0	0	0.3−0.62
	1		−50	0.035	0	0	0.35−0.65
	1		0	0.072	0	0	0.46−0.88
	1		25	−0.004	0	0	0.51−1.11
	1		50	−0.005	−0.003	0	0.98−2.60
	0		102	−0.02	−0.0007	0	5.3−7.95
	1	RbCl	25	0.029	0	0	1.06−3.0
	1		40	0.033	0	0	0.57−1.73
	1	$ZnCl_2$	25	0.035	0	0	0.8−5.0
	1		40	0.035	−0.0008	0	2.08−11.8
$LiClO_3$	1	LiCl	3	−0.015	−0.0005	0	2.0−5.47
$LiNO_3$	3	NH_4NO_3	25	0.020	0.008	0	0.23−4.99
	0		25	0.012	0	0	4.99−20
	0		31	0.012	0	0	0.45−27.04
Li_2SO_4	0	Ag_2SO_4	25	0.054	0	0	0.02−0.04
	1	$CdSO_4$	30	−0.051	0.009	0	0.18−2.45
	1	$CuSO_4$	0	−0.007	0	0	0.09−0.49
	1		25	−0.012	0	0	0.31−0.82
	1		30	−0.016	0	0	0.29−0.90
	1		55	0	0	0	0.53−1.5
	1	$FeSO_4$	30	−0.036	0	0	0.3−1.57
	1	H_2SO_4	12.5	0.018	−0.0035	0	1.3−7.55
	1		30	0.018	0	0	0.7−14.7
	1	LiCl	25	−0.067	0	0	0.3−6.37
	1		30	−0.067	0	0	0.8−16.3
	1	$LiNO_3$	25	−0.052	0.0006	0	3.7−11.18
	1		35	−0.061	0.0018	0	1.9−22
$MgCl_2$	6	$BeCl_2$	25	−0.0215	0	0	1.29−8.19

contd.

Component 1	H₂O	Component 2	t °C	Q_1	Q_1	Q_3	Data for c_2
MgCl₂	12	CaCl₂	−30	−0.06	0	0	0.9−3.12
	8		−15	−0.035	0	0	0.59−1.19
	6		0	−0.041	0	0	0.62−2.0
	6		25	−0.0315	0	0	1.49−5.17
	6		35	−0.028	0	0	0.52−3.68
	6		75	−0.010	0	0	0.58−1.29
	6	CoCl₂	25	0.029	0	0	0.43−1.57
	6	LiCl	0	−0.074	0.007	0	0.94−5.4
	6		25	−0.034	0.0018	0	1.52−5.81
	6		30	−0.036	0.0026	0	1.85−5.39
	6		70	−0.023	0.0040	0	2.0−4.29
	6		102	0.060	−0.01	0	2.07−4.26
	6	Mg(NO₃)₂	15	−0.023	0	0	0.48−1.7
	6		25	−0.022	0	0	0.32−2.09
	6		50	−0.0156	0	0	0.7−3.44
	6		75	−0.0158	0	0	0.8−4.62
	6	MgSO₄	25	−0.018	0	0	0.18−0.39
Mg(IO₃)₂	10	Mg(NO₃)₂	5	0.540	−0.1630	0	0.17−1.46
	4		5	0.346	−0.08	0	1.46−4.31
	4		25	0.24	−0.09	0.01	0.26−4.88
	4		50	0.108	−0.025	0.0015	0.58−5.7
	10	NaIO₃	5	0.34	0	0	0.07−0.13
	4		25	0.17	0.025	0	0.15−0.43
	4		50	0.135	0.012	0	0.27−0.75
MgMoO₄	5	MgCl₂	25	−0.152	0.015	0	0.25−3.88
	5	MgSO₄	25	0.115	−0.043	0	0.29−2.77
	5	Na₂MoO₄	25	0.033	0	0	0.28−1.14
Mg(NO₃)₂	6	HNO₃	25	0.005	0	0	3.6−17.4
	0	KNO₃	0	0.05	0	0	0.17−0.54
	0		25	0.03	0.003	0	0.7−1.57
	0		50	0.03	0.001	0	1.67−3.33
	0		75	0.03	0	0	2.39−6.51
	2	MgCl₂	110	0.001	0	0	4.17−6.56
	6	Mg(IO₃)₂	5	−0.004	0	0	0.02−0.10
	6		25	0	0.22	0	0.04−0.11
	6		50	−0.002	0	0	0.07−0.16
	6	MgSO₄	0	−0.05	0	0	0.13−0.18
	6		25	−0.075	0	0	0.14−0.41
	6	Mn(NO₃)₂	20	−0.007	0.010	0	0.28−3.67
	6	NaNO₃	25	0.01	0	0	0.41−2.0

contd.

Component 1	H$_2$O	Component 2	t °C	Q_1	Q_2	Q_3	Data for c_2
MgSO$_4$	7	CdSO$_4$	25	0.011	−0.0055	0	0.44−2.06
	7		40	−0.004	0.090	0	0.18−0.70
	6	CuSO$_4$	0	0.052	0	0	0.16−0.56
	7	H$_2$SO$_4$	12.6	0.012	0	0	1.22−3.89
	7	K$_2$SO$_4$	0	0.18	0	0	0.16−0.26
	7		25	0.12	0	0	0.13−0.33
	7		30	0.07	0	0	0.31−0.37
	7		35	0.07	0	0	0.14−0.37
	6		50	0.06	0	0	0.23−0.38
	6		55	0.07	0	0	0.2−0.39
	6		66	0.12	0	0	0.2−0.39
	6		85	0.06	0	0	0.2900
	7	MgCl$_2$	0	−0.170	0.020	0	0.5−5.42
	7		25	−0.107	0.005	0	0,23−3.92
	6		25	−0.085	0.004	0	3.92−5.68
	7		35	−0.085	0.004	0	0.83−2.89
	1		75	−0.073	−0.010	0	0.44−5.27
	1		100	−0.107	0.004	0	0.43−6.55
	7	MgMoO$_4$	25	−0.027	0	0	0.16−0.38
	7	Mg(NO$_3$)$_2$	0	−0.173	0.015	0	0.09−4.05
	7		25	−0.123	0.006	0	0.61−4.67
	7	MnSO$_4$	0	0.253	−0.232	0.053	0.3−2.23
	7		23	0.002	−0.010	0	0.1−3.03
	6		50	−0.006	−0.005	0	0.22−1.39
MgSeO$_4$	6	H$_2$SeO$_4$	30	0.027	0	0	0.5−5.89
	6	Na$_2$SeO$_4$	25	0	0.0025	0	0.96−2.84
MgSiF$_6$	6	H$_2$SiF$_6$	20	−0.10	0	0	0.08−5.97
	6	(NH$_4$)$_2$SiF$_6$	25	0.13	−0.03	0	0.17−0.93
MnSO$_4$	5	BeSO$_4$	25	−0.066	0.010	0	0.8−2.67
	1	H$_2$SO$_4$	25	−0.05	0.0010	0	0.03−24.27
	1		45	−0.037	0.0010	0	3.21−24.8
	1		65	−0.037	0.0010	0	3.3−25.7
	1		95	−0.028	0.0014	0	24.8−27.58
	4	K$_2$SO$_4$	25	0.068	0	0	0.04−0.41
	7	MgSO$_4$	0	−0.10	0.007	0	0.23−1.01
	5		23	−0.05	0	0	0.14−1.13
	1		50	−0.05	0	0	0.3−3.31
	4	Na$_2$SO$_4$	25	0.045	0	0	0.19−0.61
	1		35	−0.012	0	0	0.35−0.79
	1		97	−0.073	0.080	0	0.21−1.28

contd.

Component 1	H_2O	Component 2	t °C	Q_1	Q_2	Q_3	Data for c_2
$MnCl_2$	4	LiCl	0	−0.104	0.011	0	1.6−5.68
	4		20	−0.075	0.011	0	2.2−5.08
	4		35	−0.061	0.0105	0	2.2−3.81
	4		60	−0.050	0.003	0	1.86−4.11
	2		80	−0.054	0.006	0	3.4−7.46
	2		99	−0.050	0.006	0	3.8−7.45
$Mn(NO_3)_2$	6	HNO_3	20	0.014	0.006	0	0.94−3.33
	4		20	0.009	0	0	3.33−13.6
	2		20	0.021	−0.0007	0	13.6−18.14
	6	$Mg(NO_3)_2$	20	−0.02	0	0	0.9−4.02
NH_4Br	0	$CoBr_2$	0	−0.076	0.0020	0	2.26−4.18
	0		25	−0.06	0.004	0	0.2−5.06
	0		40	−0.055	0	0	0.22−5.99
	0		55	−0.055	0.005	0	0.35−7.40
	0		75	0.004	0	0	0.32−8.01
	0		150	0.004	0.003	0	1.13−7.64
	0	$PbBr_2$	25	0.161	−0.0161	0	0.11−0.44
$(NH_4)_2B_4O_7$	4	$(NH_4)_2SO_4$	10	0.27	−0.024	0	0.7−5.42
	4		35	0.20	−0.054	0.507	0.13−3.0
NH_4Cl	0	HCl	0	−0.047	0.5015	0	4.5−8.5
	0	$HgCl_2$	30	0.065	0	0	1.3−4.81
	0	KCl	0	−0.006	0	0	1.48−1.83
	0		20	−0.006	0	0	0.44−2.05
	0		25	−0.012	0	0	1.96−2.10
	0		30	−0.030	0	0	0.6−2.04
	0		45	−0.010	0	0	1.02−2.34
	0		65	−0.005	0	0	2.5211
	0		90	−0.003	0	0	2.4−2.88
	0	$MgCl_2$	25	−0.070	0.004	0	1.09−3.15
	0		60	−0.070	0.004	0	1.23−3.34
	0	$(NH_4)_2Cr_2O_7$	75	−0.028	0.005	0.05	0.1−0.81
	0	NH_4F	25	0.0125	−0.0005	0	4.1−22.1
	0	NH_4HF_2	25	0.0125	−0.0005	0	1.04−6.5
	0	$NH_4H_2PO_4$	−15	−0.008	0	0	0.3369
	0		0	0.0391	0	0	0.4046
	0	NH_4I	25	0.003	0	0	0.48−12.5
	0	NH_4NO_3	0	0.022	−0.0008	0	1.39−19.47
	0		−10	0.05	−0.0035	0	1.35−8.34
	0		10	0.034	−0.005	0	1.4−6.5
	0		20	0.020	−0.0006	0	1.4−21.34

contd.

Component 1	H₂O	Component 2	t °C	Q_1	Q_2	Q_3	Data for c_2
NH₄Cl	0	NH₄NO₃	25	0.020	−0.0005	0	1.73−24.9
			30	0.010	0	0	1.39−20.59
	0		50	0.020	−0.0005	0	2.3−10
	0	(NH₄)₂SO₄	20	0.0155	0.001	0	5.0−6.4
	0		30	0.0155	−0.0017	0	5.88−6.8
	0	(NH₄)₂SiF₆	25	0.032	0	0	0.06−0.21
	0	NaCl	−10	−0.005	−0.0033	0	2.5−4.7
	0		0	−0.005	−0.0033	0	3.0−4.7
	0		10	−0.005	−0.003	0	2.8−4.6
	0		20	−0.005	−0.003	0	1.9−4.4
	0		35	−0.005	−0.003	0	3.4−4.08
	0		50	−0.005	−0.0025	0	1.0−4.08
	0		65	−0.005	−0.0025	0	2.29−3.53
	0		80	−0.005	−0.0025	0	2.89−3.37
	0	ZnCl₂	20	0.108	−0.005	0	0.55−4.28
	0		30	0.098	−0.005	0	1.09−4.49
NH₄ClO₄	0	(NH₄)₂SO₄	25	0.034	−0.0048	0	1.1−5.7
	0		60	0.021	−0.004	0	0.82−6.08
(NH₄)₂Cr₂O₇	0	NH₄Cl	19.7	−0.01	0	0	1.03−6.96
	0		50	−0.01	−0.0006	0	1.96−9.26
	0	(NH₄)₂SO₄	50	0	0	0	0.8−7.1
	0	Na₂Cr₂O₇	0	0.037	0	0	3.7−7.0
	0		20	0.008	0	0	1.6−7.2
	0		50	−0.027	0.0028	0	3.2−9.23
	0		75	0.048	0.006	0	0.5−3.7
NH₄F	0	BeF₂	0	0.018	0.022	0	0.18−0.97
	0	KF	25	−0.0106	0	0	0.45−19.97
	0	NH₄Cl	25	0.010	−0.0003	0	1.9−5.3
NH₄HCO₃	0	(NH₄)₂SO₄	0	0.085	−0.012	0	0.53−4.9
	0		7	0.033	−0.003	0	1.25−5.11
NH₄HF₂	0	NH₄Cl	25	0.0027	0	0	1.54−3.72
NH₄H₂PO₄	0	CO(NH₂)₂	0	0.0165	−0.0006	0	4.0−11.07
	0	NH₄Cl	0	−0.050	0.005	0	0.98−4.67
	0		10	−0.061	0.0060	0	0.98−4.67
	0		20	−0.061	0.0055	0	0.98−6.23
	0		25	−0.061	0.0048	0	0.98−6.23
	0		35	−0.061	0.0050	0	1.15−6.23
	0	(NH₄)₂SO₄	0	0.06	−0.005	0	0.8−5.08
	0		10	0.055	−0.005	0	0.8−5.26
	0		20	0.045	−0.0045	0	0.83−5.65

contd.

Component 1	H$_2$O	Component 2	t °C	Q_1	Q_2	Q_3	Data for c_2
NH$_4$H$_2$PO$_4$	0	NH$_4$NO$_3$	0	0	0	0	1.39—13.5
	0		10	−0.002	0	0	1.37—17.76
	0		20	−0.015	0.001	0	1.38—16
	0		30	−0.025	0.0012	0	1.4—18.7
	0		50	−0.025	0.001	0	3.04—18.8
	0	NaH$_2$PO$_4$	−4.3	0.074	−0.008	0	0.7—4.16
	0		0	0.068	−0.0075	0	0.69—3.7
	0		10	0.060	−0.0055	0	0.7—5.5
	0		30	0.033	−0.0020	0	0.68—7.79
NH$_4$HSO$_3$	0	(NH$_4$)$_2$SO$_4$	−31	0.017	−0.0055	0	0.51—1.94
	0		−19	0.01	−0.008	0	0.63—1.95
	0		−10	0.01	−0.008	0	0.7—1.94
	0		0	0.01	−0.008	0	0.55—2.05
	0		20	0.01	−0.0023	0	2.0452
NH$_4$I	0	NH$_4$Cl	25	−0.005	0	0	3.77
NH$_4$NO$_3$	0	AgNO$_3$	30	0.027	−0.001	0	3.08—9.87
	0	Ca(NO$_3$)$_2$	0	−0.005	0.0020	0	1.09—8.76
	0		10	−0.014	0.0030	0	1.2—10.53
	0		20	−0.021	0.0029	0	1.3—10.44
	0		25	−0.015	0.0019	0	0.92—12.74
	0		30	−0.015	0.0019	0	1.26—19.21
	0		40	−0.015	0.0026	0	2.32—10.81
	0	Cu(NO$_3$)$_2$	30	−0.058	0.006	0	1.7—8.6
	0	HNO$_3$	0	−0.018	0.0014	0	3.0—20
	0		15	−0.018	0.0014	0	0.2—15
	0		20	−0.018	0.0014	0	0—15
	0	NH$_4$Cl	0.4	−0.006	0	0	1.37—5.54
	0		25	−0.010	0.0025	0	2.27—5.78
	0		50	−0.010	0.0025	0	3.0—6.0
	0	(NH$_4$)$_2$SO$_4$	0	−0.005	0.003	0	0.54—1.75
	0		25	−0.005	0.0025	0	0.5—1.12
	0		30	−0.005	0.0025	0	0.6—0.88
	0		40	−0.005	0.0100	0	0.5—0.9
	0		70	−0.005	−0.01	0	1.12
	0	NaNO$_3$	−17	0.0125	0	0	2.0—6.7
	0		−10	0.010	0	0	2.3—7.16
	0		0	0.005	0	0	0.5—6.1
	0		20	0.005	0	0	0.92—6.19
	0		30	0.004	0	0	1.0—7.5
	0		60	0.0055	0	0	1.37—15.4
	0		80	0.007	0	0	9.2—26.06
	0		98	0.008	0	0	3.9—20
	0	Pb(NO$_3$)$_2$	20	0.022	0	0	0.9—3.89
	0		25	0.022	−0.001	0	2.1—4.12

contd.

Component 1	H$_2$O	Component 2	t °C	Q_1	Q_2	Q_3	Data for c_2
NH$_4$NO$_3$	0	UO$_2$(NO$_3$)$_2$	25	−0.007	−0.001	0	0.5 — 4.15
NH$_4$SCN	0	AgSCN	25	0.053	−0.0012	0	0.57 — 0.82
	0	Ba(SCN)$_2$	25	−0.023	0.003	0	1.33 — 9.35
(NH$_4$)$_2$SO$_3$	1	NH$_4$HSO$_3$	0	0.0138	0	0	3.3 — 20.31
	0		0	0	0.0046	0	2.7 — 3.73
	0		10	0.0115	0	0	0.88 — 3.85
	0		15	0.0115	0	0	0.44 — 3.87
	1		20	0.011	0	0	2.5 — 27.6
	0		20	0.011	0.0008	0	2.2 — 3.8
	1		30	0.011	0	0	3.7 — 32.8
	0		30	0.0105	0	0	0.3 — 3.8
(NH$_4$)$_2$SO$_4$	0	BeSO$_4$	0	−0.015	0.0037	0	0.84 — 5.11
	0		25	0.006	0.0030	0	1.55 — 4.39
	0		50	0.0215	0	0	1.01 — 4.72
	0		60	0.0215	0	0	2.32 — 4.53
	0		75	0.027	0	0	2.09 — 4.19
	0		99.5	0.0315	0	0	1.31 — 3.32
	0	CdSO$_4$	25	0.006	0.006	0	0.20 — 0.68
	0		50	−0.001	0.0067	0	0.42 — 1.78
	0		77	0.031	0	0	0.53 — 1.37
	0		97	0.025	0	0	0.47 — 1.05
	0	H$_2$SO$_4$	30	0.046	0.0014	0	2.0 — 3.2
	0		50	0.046	0.0012	0	4.21
	0		98.5	0.046	0.0004	0	2.0 — 8.25
	0	Li$_2$SO$_4$	30	−0.005	0	0	0.48 — 1.11
	0		71	0.01	0	0	0.9 — 1.01
	0		95	−0.008	0	0	0.57 — 1.14
	0	NH$_4$Cl	20	−0.0036	−0.0005	0	1.85 — 5.04
	0		30	−0.0036	−0.0006	0	3.6 — 5.88
	0		30	0.006	−0.0015	0	0.91 — 6.03
	0	(NH$_4$)$_2$B$_4$O$_7$	10	0.015	−0.10	0	0.1459
	0		20	0.015	−0.05	0	0.2193
	0		25	0.015	−0.001	0	0.06 — 0.26
	0		35	0.015	−0.005	0	0.16 — 0.40
	0		50	0.015	0.015	0	0.8942
	0	(NH$_4$)$_2$Cr$_2$O$_7$	5	0.107	−0.01	0	0.8 — 5.7
	0		25	0.046	−0.0043	0	0.8 — 6.37
	0		50	0.170	−0.010	0	0.03 — 0.4
	0	NH$_4$H$_2$PO$_4$	−10	0.021	−0.023	0	0.3 — 0.69
	0		−5	0.021	−0.0125	0	0.3 — 0.79
	0		0	0.021	−0.013	0	0.3 — 0.82
	0		10	0.021	0	0	0.3 — 1.09
	0		20	0.021	0	0	0.3 — 1.35
	0		30	0.021	0	0	0.3 — 1.72

contd.

Component 1	H$_2$O	Component 2	t °C	Q_1	Q_2	Q_3	Data for c_2
(NH$_4$)$_2$SO$_4$	0	NH$_4$HSO$_3$	−19	0.005	0	0	1.79−16.89
	0		0	0.005	0	0	1.79−24.13
	0		10	0.005	0	0	1.8−19.66
	0		20	0.005	0	0	1.7−32.62
	0	NH$_4$NO$_3$	0	0.034	0	0	1.2−12.84
	0		25	0.0035	0	0	3.6−15.57
	0		30	0.0038	0	0	2.8−16.8
	0		40	0.0038	0	0	3.49−20.27
	0		70	0.0038	0	0	2.8−37.87
	0	(NH$_4$)$_2$SO$_3$	−13	0	0	0	0.44−1.85
	0		−10	0	0	0	0.44−1.93
	0		0	0.0038	0	0	0.45−1.95
	0		10	0.0038	0	0	0.44−2.94
	0		15	0.004	0	0	0.83−3.19
	0		20	0.0048	0	0	0.43−3.61
	0		43	0.0058	0	0	0.42−4.23
	0	NH$_4$SO$_3$. NH$_2$	0	0.004	0	0	3.4−9.72
	0		25	0.004	0	0	17.603
	0		30	0.004	0	0	3.8−19.44
	0		50	0.005	0	0	4.2−30.5
	0		70	0.005	0	0	9.56−45.5
NH$_4$SO$_3$.NH$_2$	0	(NH$_4$)$_2$SO$_4$	0	0.017	−0.0023	0	1.3−3.5
	0		25	0.017	−0.0030	0	2.5698
	0		30	0.017	−0.0030	0	0.68−2.46
	0		50	0.017	−0.0030	0	2.05−2.63
	0		70	0.017	−0.0040	0	1.78−2.57
(NH$_4$)$_2$SeO$_4$	0	CaSeO$_4$	30	0.002	0	0	0.0438−0.26
(NH$_4$)$_2$SiF$_6$	0	H$_2$SiF$_6$	25	0.044	−0.016	0	0.6−2.96
	0	NH$_4$F	5	0.034	0	0	2.66−7.07
	0	MgSiF$_6$	25	0.048	−0.019	0	0.61−1.81
Na$_2$B$_4$O$_7$	10	(NH$_4$)$_2$B$_4$O$_7$	25	0.54	−0.16	0	0.35−0.55
	10		30	0.463	−0.0885	0	0.52−0.69
	10	Na$_2$CO$_3$	25	0.928	−0.605	−0.25	0.2−3.0
	10		35	0.445	−0.073	0	0.33−2.53
	10		50	0.263	−0.028	0	0.5−3.2
	10		55	0.170	0	0	0.35−0.5
	10		55	0.226	−0.022	0	0.9−2.4
	4		55	0.226	−0.031	0	2.4−3.3
	4		60	0.270	−0.073	0.0075	0.25−3.36
	10	NaF	25	−0.068	0.19	0	0.2−0.9
	10	NaHCO$_3$	25	0.49	−0.13	0	0.2−1.3
	10		50	0.15	0	0	0.2−1.7

contd.

Component 1	H_2O	Component 2	t °C	Q_1	Q_2	Q_3	Data for c_2
$Na_2B_4O_7$	10	Na_2SO_4	10	1.73	-2.6	1.7	$0.13-0.64$
	10		20	0.94	-0.64	0.19	$0.72-1.36$
	10		25	0.90	-0.64	0	$0-2.0$
	10		30	0.785	-0.19	0	2.9
	10		45	0.305	-0.044	0	$0.33-3.2$
	10		55	0.1818	0	0	$1.17-1.6$
	5		55	0.295	-0.121	0.02	$1.6-3.0$
	4		65	0.156	-0.069	0.016	$0.32-2.8$
NaBr	2	HBr	25	-0.025	-0.002	0	$0-9.0$
	0		25	-0.041	-0.0013	0	$9.0-16$
	2		44.5	-0.025	0	0	$1.0-2.6$
	0		44.5	-0.033	0	0	$2.8-5.3$
	0		65	-0.056	0	0	$1.0-5.4$
	0	KBr	-20	-0.010	0	0	$0.3-0.5$
	0		-10	-0.010	0	0	$0.3-0.7$
	0		0	0.010	0	0	$0.3-0.75$
	0		10	0.010	0	0	$0.37-0.9$
	0		20	0.010	0	0	$0.4-1.1$
	0		25	0.010	0	0	$0.4-1.15$
	0		30	0.010	0	0	$0.4-0.9$
	0		35	0.010	0	0	$1.3-1.6$
	2	Na_2CO_3	30	-0.041	0	0	$0.47-0.54$
	0		80	-0.041	0	0	$0.7-9.5$
	2	$NaClO_3$	25	0.015	0	0	$1.4-2.9$
	2	NaI	25	-0.005	-0.0015	0	$0.5-10.3$
	2	$NaNO_3$	25	0.023	-0.002	0	$2.0-3.5$
	2	NaOH	25	-0.02	0.0025	0	$0.4-5.25$
	0		25	-0.026	0.0006	0	$5.6-15$
	2		44.5	-0.008	0	0	$0.88-2.0$
	0		44.5	-0.020	0	0	$3.0-8.0$
	0		65	-0.020	0.0002	0	$0-11$
	2		17	-0.055	0.011	0	$0.81-3.35$
	0		17	-0.01	0	0	$4.29-13.63$
	0	$PbBr_2$	25	0.092	-0.099	0	$1.3-3.6$
$NaBrO_3$	0	NaBr	10	-0.020	0.001	0	$0.68-8.17$
	0		25	-0.023	0.0009	0	$0-10$
	0		35	-0.038	0.002	0	$1.0-10$
	0		45	-0.023	0	0	$1.0-11$
	0	Na_2CO_3	80	0.013	-0.008	0	$0.9-3.0$
	0	NaCl	25	-0.021	0	0	$0.9-6.0$
	0	$NaClO_3$	25	0.01	-0.001	0	$2.7-8.0$

contd.

Component 1	H$_2$O	Component 2	t °C	Q_1	Q_2	Q_3	Data for c_2
NaBrO$_3$	0	NaHCO$_3$	35	−0.094	0.11	0	0.3−0.9
	0	NaI	25	−0.025	0.0006	0	1.74−12.25
	0	Na$_2$MoO$_4$	25	−0.082	0.435	−0.065	1.1−3.14
	0	NaNO$_3$	25	0.026	−0.0011	0	0.8−10.5
	0	Na$_2$SO$_4$	10	0.09	−0.057	0	0−0.4
	0		25	0.043	−0.165	0	0.4−1.7
	0		45	0.0185	−0.0075	0	0.9−2.2
NaCN	2	NaOH	25	0.13	0	0	1.0−43
	0		25	−0.014	0	0	4.54−27.6
	0		35	−0.013	0.0004	0	3.75−19.9
	0		55	−0.007	0	0	3.5−31.0
Na$_2$CO$_3$	10	Na$_2$B$_4$O$_7$	25	0.166	0.73	0	0.06−0.16
	10	NaBr	30	0.031	0.0055	0	0.33−0.9
	7		30	−0.022	0.0025	0	0.9−2.59
			30	−0.005	0.0015	0	2.59−8.51
	1	NaBrO$_3$	80	−0.02	0	0	0.9−2.5
	10	NaCl	0	0.03	0.018	−0.0022	0.8−5.7
	10		15	0.025	0.003	0	2.1−5.0
	10		20	0.014	0.0063	0	0.9−4.3
	10		25	0.050	0.0058	0	3.0−4.0
	10		30	0.022	0	0	0.91−0.96
	7		30	−0.027	0.007	0	1.0−2.6
	7		35	0	0	0	0−0.7
	0		30	−0.044	0.003	0	2.6−3.8
	0		40	−0.040	0.0023	0	1.0−4.4
	0		60	−0.030	0	0	1.8−4.5
	10	NaHCO$_3$	0	0.52	0	0	0.1−0.6
	10		15	0.52	0	0	0.2−0.6
	10		20	0.52	0	0	0.4−0.6
	10	NaI	30	−0.065	0.08	0	0.22−0.96
	7		30	−0.056	0.0076	0	0.95−2.85
	1		30	−0.063	0.0017	0	2.85−12.83
	10	NaIO$_3$	25	−0.047	0.97	0	0.03−0.14
	1		40	−0.047	0.97	2.7	0.03−0.13
	1		50	−0.047	0.35	0	0.09−0.18
	10	NaNO$_2$	23.1	0.032	0	0	2.0−7.5
	1		23.1	−0.015	0	0	7.5−10
	10	NaOH	0	0.0375	0.0033	0	2.2−7.47
	1		0	−0.076	0.0024	0	7.78−10.33
	7		0	−0.0345	0.003	0	7.47−7.78
	10		15	−0.0135	0.0094	0	1.4−5.73

contd.

Component 1	H$_2$O	Component 2	t °C	Q_1	Q_2	Q_3	Data for c_2
Na$_2$CO$_3$	1	NaOH	15	−0.015	0	0	6.67−22.79
	10		20	−0.0135	0.0089	0	1.94−4.51
	1		20	−0.015	0	0	5.57−17.82
	10		25	−0.0135	0.006	0	1.76−3.19
	7		25	−0.019	0	0	3.19−4.41
	1		25	−0.025	0.0005	0	4.41−18.56
	10		30	−0.0135	0.029	0	1.1−1.25
	7		30	−0.013	0	0	1.25−2.81
	1		30	−0.039	0	0	2.81−11.48
	1		35	−0.042	0	0	1.75−16.25
	1		45	−0.034	−0.0022	0	4.12−15.64
	1		60	−0.039	0	0	2.13−13.27
	1		100	−0.042	0.0015	0	0.7−10.98
	10	Na$_2$S	30	−0.0365	−0.0075	0	0.5−1.8
	1		60	−0.196	0.209	−0.06	0.47−1.26
	1		80	−0.196	0.244	−0.06	0.29−0.81
Na$_2$C$_2$O$_4$	2	NaIO$_3$	40	0.084	0	0	0.31−0.53
	2		50	0.088	0	0	0.53−0.68
	2		60	0.19	−0.032	0	0.8−3.13
NaCl	0	BeCl$_2$	25	−0.07	−0.021	0	0.6−4.4
	2	CaCl$_2$	−45	−0.015	−0.029	0	3.5−4.0
	2		−40	−0.015	−0.02	0	3.33−4.14
	2		−30	−0.015	−0.020	0	2.4−4.32
	2		−35	−0.015	−0.020	0	2.94−4.25
	2		−25	−0.015	−0.02	0	1.34−4.49
	2		−20	−0.015	−0.020	0	1.58−3.84
	0		−20	−0.015	−0.021	0	3.84−4.58
	2		−15	−0.015	−0.02	0	1.60−3.35
	0		−15	−0.015	−0.02	0	3.34−4.66
	2		−10	−0.015	−0.02	0	1.66−3.08
	0		−10	−0.015	−0.022	0	3.08−4.68
	0		−5	−0.015	−0.02	0	0.58−4.7
	0		0	−0.057	−0.010	0	1.17−5.5
	0		25	−0.057	−0.005	0	3.61−7.06
	0		50	−0.057	−0.005	0	0.41−11.85
	0		94.5	−0.04	0	0	1.47−12.73
	0	CdCl$_2$	100	0.098	−0.0083	0	1.76−3.34
	0	CoCl$_2$	20	−0.005	−0.007	0	1.18−5.47
	0		25	−0.005	−0.007	0	0.72−5.53
	0		38	−0.03	0	0	1.31−6.87
	0		50	0	−0.006	0	0.72−1.95
	0		100	0.004	0	0	1.51−11.84
	0	CuCl	26.5	0.131	−0.015	0	1.88−4.45
	0	CsCl	25	0.0065	0	0	0.7−5.5
	0		50	0.0055	0	0	0.7−8.3
	0		75	0.0045	0	0	1.0−10.5
	0	CuCl$_2$	30	0.010	0	0	1.2−5.18

contd.

Component 1	H_2O	Component 2	t °C	Q_1	Q_2	Q_3	Data for c_2
NaCl	0	HCl	0	−0.113	0.003	0	8.0−22
	0		25	0.008	−0.0019	0	1.4−15.5
	0		30	−0.047	−0.0012	0	2.47−15.2
	0	KCl	0	0	0.0006	0	0.4−1.4
	0		20	0	0.0007	0	0.65−1.95
	0		25	0	−0.0010	0	0.6−2.18
	0		40	0	0	0	0.8−2.62
	0		60	0	0	0	1.0−3.3
	0		80	0	0	0	1.4−4.0
	0		100	0	0	0	2.0−4.75
	0		120	0	0.0015	0	1.9−5.5
	0		140	0.0025	0	0	2.0−6.2
	0		150	0.0025	0	0	4.0−6.5
	0		169.5	0.0025	0	0	1.78−7.3
	0		189.5	0.0025	0.0002	0	2.0−8.2
	0		200	0.005	0	0	4.0−8.7
	0	KNO_3	20	0.014	−0.0021	0	0.9−3.7
	0		30	0.016	−0.0022	0	1.2−4.9
	0		40	0.013	−0.0012	0	1.7−6.4
	0		91	0.013	−0.0007	0	5.6−21.65
	0	$MgCl_2$	15	−0.058	0	0	0.1−3.15
	0		25	−0.058	−0.0105	0	0.07−5.74
	0		55	−0.049	−0.0100	0	1.05−6.33
	0		83	−0.030	−0.011	0	1.15−6.7
	0		105	−0.030	−0.011	0	3.8−7.91
	0	NH_4Cl	−10	0.003	0	0	2.2805
	0		0	0.003	−0.001	0	2.6768
	0		10	0.003	−0.001	0	2.5−3.43
	0		20	0.003	−0.0005	0	0.5−4.01
	0		50	0.003	0	0	1.0−6.16
	0		65	0.003	0	0	5.7−7.98
	0		80	0.003	0	0	0.3−9.47
	0	$NaBrO_3$	25	0.015	0	0	0.2−0.7
	0	$NaClO_3$	−19.2	0.0105	0	0	1.6−3.1
	0		−9.8	0.0105	0	0	0.9−3.4
	0		10	0.004	−0.0008	0	1.2−4.7
	0		20	−0.017	0.002	0	1.8−4.8
	0		30	0.003	0.0006	0	0.7−6.6
	0		50	0	0	0	3.9−9.0
	0		70	0	0	0	1.9−11.6
	0		100	−0.001	0	0	2.2−17.3
	0	$NaClO_4$	10	−0.005	−0.0007	0	3.2−12.9
	0		25	−0.013	0	0	6.8−26.5
	0	Na_2CO_3	15	0.009	−0.009	0	0.4−1.2
	0		20	0.009	−0.009	0	1.5−1.8
	0		25	0	0	0	1.0−2.4
	0		30	−0.003	0	0	0.6−2.5
	0		35	0	0	0	0.5−2.4
	0		40	0	0	0	0.5−2.1

contd.

Component 1	H$_2$O	Component 2	t °C	Q_1	Q_2	Q_3	Data for c_2
NaCl	0	Na$_2$Cr$_2$O$_7$	0	0.01	−0.008	0	0−6.0
	0		20	0.005	−0.006	0	3.0−7.0
	0		25	−0.017	−0.004	0	0−7.0
	0		50	−0.027	0	0	0−9.0
	0		75	−0.015	0	0	0−13
	0		100	−0.028	0	0	0−16
	0	Na$_2$HPO$_4$	25	−0.136	0.120	0	0.3−0.85
	0	NaI	0	−0.026	−0.0018	0	0.5−10.5
	0		10	−0.026	−0.0018	0	1.0−11
	0		25	−0.026	−0.0018	0	1.0−12
	0		50	−0.041	0	0	2.0−15
	0		75	−0.030	0	0	2.0−20
	0	NaIO$_3$	35	0.35	0	0	0.11−0.16
	0	NaNO$_2$	25	0	0	0	2.0−10
	0		45	−0.002	0	0	1.3−12.4
	0	NaNO$_3$	0	0.015	−0.0012	0	1.4−4.6
	0		20	0.01	−0.006	0	3.5−6.4
	0		25	0.015	−0.0014	0	1.7−7.0
	0		30	0.012	−0.009	0	2.8−7.5
	0		40	0.012	−0.009	0	3.0−8.7
	0		50	0.015	−0.0013	0	2.8−10
	0		75	0.006	0	0	5.4−14
	0		100	0.006	−0.0003	0	4.8−18.3
	0	NaOH	50	−0.01	0	0	1.7−31.5
	0		60	−0.01	0	0	1.7−25.9
	0		70	−0.01	0	0	1.7−21.3
	0		80	−0.008	0	0	1.7−21.5
	0		90	−0.008	0	0	1.75−21.63
	0	Na$_2$SO$_3$	25	−0.003	0	0	0.1−1.35
	0		40	−0.002	0	0	0.3−0.5
	0		60	−0.0015	0	0	0.2−0.35
	0		80	−0.0017	0	0	0.1−0.3
	0		100	0	0	0	0.1−0.3
	0	Na$_2$SO$_3$.NH$_2$	25	0.023	0	0	1.3−6.6
	0	Na$_2$SO$_4$	38	−0.015	0.040	0	0.1−0.6
	0		50	−0.010	0.044	0	0.2−0.6
	0		75	−0.011	0.048	0	0.2−0.5
	0		100	−0.005	0.08	−0.093	0.2−0.45
	0	Na$_2$S$_2$O$_3$	25	0.006	−0.005	0	0.4−3.6
	0	PbCl$_2$	13	0.18	−0.48	1.5	0.03−0.09
	0		50	0.18	−0.44	1.1	0−0.25
	0		100	0.105	0	0	0.1−0.7
	0	SrCl$_2$	18	−0.028	−0.0128	0	0.8−2.7
	0		60	−0.023	−0.013	0	0.8−2.4

contd.

Component 1	H$_2$O	Component 2	t °C	Q_1	Q_2	Q_3	Data for c_2
NaCl	0	SrCl$_2$	60	−0.056	0	0	4.0−5.3
	0		100	−0.038	0	0	0.8−6.3
NaClO$_3$	0	KClO$_3$	10	0.034	0	0	0.1−0.19
	0		30	0.032	0	0	0.16−0.42
	0		50	0.0275	0	0	0.52−0.81
	0		70	0.021	0	0	0.63−1.47
	0		100	0.017	0	0	1.36−3.31
	0	NaBr	25	−0.027	0	0	2.3−8.6
	0	Na$_2$CO$_3$	24.2	0.007	−0.014	0	0.5−2.0
	0	NaCl	−9.8	−0.024	0	0	1.5−4.6
	0		10	−0.026	0	0	0.9−4.3
	0		20	−0.038	0	0	0.8−3.9
	0		30	−0.022	0	0	0.6−3.6
	0		50	−0.018	0	0	0.9−3.1
	0		70	−0.016	0	0	0.8−2.6
	0	NaI	25	−0.035	0	0	1.7−12.1
	0	NaIO$_3$	25	0.010	0	0	2.3−7.7
	0		50	0.012	0	0	0.14−0.19
	0	NaNO$_3$	25	0.010	0	0	2.3−7.7
	0	NaOH	18	−0.022	0	0	3.0−14
	0	Na$_2$SO$_4$	15	−0.01	0	0	0.3−0.6
	0		18	−0.018	0.0332	0	0.3−0.7
	0		25	−0.005	0	0	0.4−0.5
	0		45	−0.01	0	0	0.3−0.4
NaClO$_4$	0	NaNO$_3$	25	0	0	0	0.8−4.4
Na$_2$CrO$_4$	4	NaClO$_3$	19	0.013	0	0	0.4−2.8
	4		25	0.013	−0.0017	0	1.3−2.8
	4		50	0.018	0	0	1.25−3.0
Na$_2$Cr$_2$O$_7$	0	K$_2$Cr$_2$O$_7$	25	0.0148	0	0	0.5352
	0		50	−0.0014	0	0	0.9356
	0		100	−0.0029	0	0	1.7601
NaF	0	HF	20	0.09	0.395	0.9	0.1−0.3
	0		40	0.205	0	0	0.2−0.6
	0	Na$_2$B$_4$O$_7$	25	−0.44	0.78	0	0−0.15
	0	NaCl	25	−0.015	0	0	6.0
	0		35	−0.015	0	0	1.0−6.0
NaHCO$_3$	0	KHCO$_3$	20	0.080	−0.0115	0	0.74−2.73
	0		25	0.080	−0.0105	0	0.48−3.44
	0		30	0.067	−0.0063	0	0.7−3.75
	0		50	0.036	0.0027	0	1.37−5.42

contd.

Component 1	H_2O	Component 2	t °C	Q_1	Q_2	Q_3	Data for c_2
$NaHCO_3$	0	NH_4HCO_3	35	0.066	−0.0072	0	0.5−3.7
	0	$Na_2B_4O_7$	25	0.335	0	0	0−0.15
	0		50	0.192	0	0	0.15−0.4
	0	NaBr	35	−0.026	0.001	0	1.3−9.5
	0	$NaBrO_3$	25	0.035	0	0	0.5−2.3
	0		35	0.027	0	0	0.5−2.8
	0	Na_2CO_3	0	0.25	−0.1344	0	0.4−0.58
	0		15	0.17	−0.0435	0	0.6−1.5
	0		20	0.138	−0.002	0	0.4−0.7
	0		30	0.114	−0.020	0	0.1−2.1
	0		35	0.105	−0.0228	0	0.1−2.1
	0		45	0.075	−0.0105	0	0.1−2.1
	0		50	0.075	−0.002	0	0.1−2.1
	0		60	0.075	−0.012	0	0.1−2.1
	0	NaCl	0	0.020	−0.0041	0	1.7−6.0
	0		15	0.007	−0.0022	0	1.7−6.1
	0		20	0.003	−0.002	0	1.6−6.1
	0		25	−0.008	0	0	0.3−6.1
	0		30	−0.010	0	0	1.7−6.1
	0		35	−0.013	0	0	2.1−6.2
	0		45	−0.017	0	0	2.1−6.2
	0		60	−0.015	0	0	2.1−6.3
	0	$NaNO_2$	23	0.002	−0.0005	0	2.2−12.3
	0	Na_2SO_4	25	0.127	−0.0275	0	0.8−1.9
	0		50	0.123	−0.030	0	0.9−3.2
NaH_2PO_4	2	$NH_4H_2PO_4$	−7	0.015	0.005	0	0.45−1.23
	2		−4.3	0.020	0.005	0	0.44−1.30
	2		0	0.025	0	0	0.36−1.45
	2		10	0.045	−0.0098	0	0.45−1.8
	2		20	0.039	−0.006	0	0.45−2.22
	2		30	0	0.001	0	0.44−2.58
	2	$NaNO_3$	−9.9	−0.0175	0	0	0.7−6.5
	2		0	−0.018	0	0	0.7−6.9
	2		10	−0.017	0	0	0.75−7.2
	2		20	−0.013	0	0	0.7−7.5
	2		30	−0.005	−0.0012	0	0.7−6.8
	2	Na_2SO_4	25	0	0	0	0.21−1.4
Na_2HPO_4	12	NaCl	25	0.14	−0.011	0	1.0−4.8
	12	Na_2SO_4	25	0.14	0	0	0.2−2.71
NaI	2	AgI	25	0.034	−0.001	0	1.32−5.15
	2	KI	8	0.01	−0.0025	0	0.97−1.26
	2		25	0.01	0	0	0.37−1.38
	2		40	0.01	0.0018	0	0.92−1.43

contd.

Component 1	H_2O	Component 2	t °C	Q_1	Q_2	Q_3	Data for c_2
NaI	2	$NaClO_3$	25	0.01	0	0	0.3—1.2
	2	$NaNO_3$	25	0.016	0	0	0.9—1.9
$NaIO_3$	5	KIO_3	5	0.383	—0.1	0	0.15—0.23
	5		25	0.285	—0.1	0	0.2—0.36
	5		50	0.212	—0.11	0	0.23—0.63
	5	$Mg(IO_3)_2$	5	0.90	0	0	0.06—0.17
	1		25	0.55	0	0	0.03—0.18
	1		50	0.70	—1.6	0	0.04—0.25
	1	Na_2CO_3	25	0.136	—0.013	0	2.2—2.8
	1		40	0.105	—0.013	0	4.1—4.6
	1		50	0.101	—0.013	0	3.3—4.5
	1	$NaClO_3$	25	0.075	—0.004	0	0.9—9.4
	1		50	0.041	—0.0018	0	1.1—4.9
	1	Na_2MoO_4	25	0.032	—0.014	0	0.3—3.0
	1	$NaNO_3$	25	0.093	—0.0044	0	0.4—10.8
	1		50	0.043	—0.0004	0	0.8—8.9
	1	Na_2SO_4	25	0.19	—0.0295	0	1.0—2.0
Na_2MoO_4	2	$NaBrO_3$	25	0.050	—0.046	0	0—0.25
	2	$NaClO_3$	25	0	0	0	0.7—7.6
	2	$NaIO_3$	25	0	0	0	0.1—0.5
	2	$NaNO_3$	25	0.01	—0.0006	0	1.6—8.3
	2	Na_2SO_4	20	0.065	—0.056	0	0.2—0.8
	2		25	0.017	—0.002	0	0.7—1.3
	2		35	0.023	—0.0075	0	0.4—1.1
$NaNO_2$	0	Na_2CO_3	23.1	—0.006	0	0	0.4—1.12
	0	NaCl	25	0	0	0	1.1—10
	0		45	—0.002	0	0	1.3—12.4
$NaNO_3$	0	$Ba(NO_3)_2$	30	0	0	0	0.07—0.16
	0	$Ca(NO_3)_2$	0	—0.020	0	0	0.35—5.95
	0		20	—0.017	0	0	0.6—7.92
	0		25	—0.023	0.0018	0	1.8—9.13
	0		50	—0.030	0.0019	0	3.21—21.68
	0		94.5	—0.018	0.0013	0	3.5—26.56
	0	HNO_3	0	—0.028	0.0005	0	5.5—45
	0		15	—0.035	0.0005	0	2.25—32
	0		20	—0.0308	0.0005	0	5.25—43
	0		75	—0.018	0.0003	0	6.2—34

contd.

Component 1	H$_2$O	Component 2	t °C	Q_1	Q_2	Q_3	Data for c_2
NaNO$_3$	0	KClO$_4$	25	0.049	−0.011	0	0.25−0.52
	0		50	0.041	−0.007	0	0.29−0.95
	0		75	0.020	0.0015	0	0.5−1.68
	0		100	0.020	0	0	1.1−2.78
	0	KNO$_3$	0	0.032	0	0	2.03
	0		20	0.032	0	0	3.75−4.14
	0		25	0.032	−0.0022	0	4.6527
	0		30	0.032	−0.0022	0	1.41−5.37
	0		40	0.028	−0.0016	0	2.0−7.31
	0		50	0.025	−0.0012	0	3.0−9.13
	0		62.5	0.019	−0.0004	0	12.56
	0		75	0.017	−0.0004	0	5.39−11.44
	0	Mg(NO$_3$)$_2$	25	−0.054	−0.008	0	0.5−3.88
	0	NH$_4$NO$_3$	−17	0.022	−0.0007	0	2.2−8.25
	0		−10	0.022	−0.0007	0	2.2−10.07
	0		0	0.022	0	0	13.19
	0		25	0.022	0	0	3.11−21.81
	0		30	0.022	0	0	2.2−10.32
	0		40	0.022	0	0	2.2−12.2
	0	NaBr	25	−0.028	0	0	3.2−8.8
	0	NaBrO$_3$	25	0.008	0	0	0.4−1.5
	0	NaCl	0	−0.025	0	0	2.7−4.9
	0		25	−0.020	0	0	1.0−4.2
	0		30	−0.022	0	0	1.6−4.0
	0		40	−0.019	0	0	1.8−3.7
	0		50	−0.017	0	0	1.3−3.6
	0		75	−0.015	0	0	1.5−3.0
	0		91	−0.008	0	0	1.5−2.7
	0		100	−0.008	0	0	1.3−2.6
	0	NaClO$_3$	25	0	0	0	1.3−8.1
	0	NaClO$_4$	25	0	0	0	1.5−13.2
	0	Na$_2$CrO$_4$	25	−0.0375	0	0	2.2−6.0
	0		50	−0.030	0	0	5.1−7.2
	0		98.5	0.011	−0.004	0	3.6−6.2
	0	NaH$_2$PO$_4$	−17.5	−0.072	0.018	0	0.5−1.0
	0		−14	−0.078	0.048	−0.019	0.5−1.1
	0		−9.9	−0.06	0.042	−0.025	0.2−1.3
	0		0	−0.03	0	0	0.6−1.5
	0		10	−0.025	0	0	0.6−2.1
	0		20	−0.019	0	0	0.7−3.0
	0		30	−0.005	−0.0035	0	0.8−4.6
	0	Na$_2$HPO$_4$	25	−0.084	0.55	0	0.1−0.7
	0	NaI	25	−0.0395	0.0008	0	4.3−12.2

contd.

Component 1	H_2O	Component 2	t °C	Q_1	Q_2	Q_3	Data for c_2
$NaNO_3$	0	$NaIO_3$	25	0.036	−0.06	0	0.1−0.25
	0		50	0.028	−0.011	−0.025	0.2−0.3
	0	Na_2MoO_4	25	−0.037	0	0	0.5−1.4
	0	$Na_2S_2O_3$	9	−0.015	−0.01	0	1.4−2.7
	0		25	−0.040	0.0016	0	1.5−4.2
	0	$Pb(NO_3)_2$	25	0.033	0	0	0.25−1.05
	0		50	0.027	0	0	0.43−1.33
	0	$UO_2(NO_3)_2$	25	−0.068	0	0	0.5−2.8
NaOH	1	NaCN	25	−0.018	0.0045	0	1.6−6.6
	1		35	−0.027	0.011	0	1.0−1.7
	1		35	0.013	0	0	3.2−5.7
	1		55	−0.019	0.0058	0	1.8−5.3
Na_2S	9	Na_2CO_3	30	−0.1	0	0	0.2−2.1
	6		60	−0.1	0	0	0.2−1.9
	6		80	−0.1	0	0	0.2−3.5
	9	Na_2SO_4	25	0.108	−0.077	0	0.3−0.6
	9		31	0.120	−0.20	0	0.1−0.4
NaSCN	0	AgSCN	25	0.039	0	0	0.80−2.23
	2	$Ba(SCN)_2$	25	0.012	0	0	0.71−4.16
	2	$Ca(SCN)_2$	25	−0.006	0	0	1.12−3.21
	0		25	−0.023	−0.0015	0	1.75−9.78
Na_2SO_3	7	NaCl	0	0.012	0.002	0	1.2−5.8
	7		25	−0.026	0.007	0	0.6−4.8
	0		40	−0.041	0.002	0	1.1−5.6
	0		60	−0.041	0.001	0	2.3−6.0
	0		80	−0.032	0	0	2.3−6.3
	0		100	−0.038	0.0017	0	2.3−6.5
	0	NaOH	20	−0.061	0.024	−0.0025	0.5−4.5
	7		20	−0.0365	0	0	4.4−14.3
	0		25	−0.018	0.0022	0	0.2−3.13
	7		25	−0.0275	−0.0018	0	3.1−18.7
	0		32	−0.0275	−0.005	0	0.2−26.9
	7	$Na_2S_2O_5$	25	0.026	0.005	0	0.2−2.6
	7		35	−0.056	0.050	−0.013	0.1−3.1
Na_2SO_4	10	Ag_2SO_4	20	0.41	0	0	0.02−0.04
	10		25	0.30	0.7	0	0.01−0.046
	10		31.5	0.185	0	0	0.014−0.031
	10	$BeSO_4$	0	0.380	−0.18	0.025	0.12−1.4
	10		25	0.085	−0.0012	0	0.59−2.35
	0		50	0.056	−0.0340	0.0055	0.82−3.93
	0		75	0.040	−0.0027	0	1.10−3.00

contd.

Component 1	H₂O	Component 2	t °C	Q_1	Q_2	Q_3	Data for c_2
Na₂SO₄	0	BeSO₄	86	0.052	−0.0030	0	1.29−3.17
	0		90.5	0.063	−0.003	0	1.59−2.22
	10	CoSO₄	25	0.099	−0.022	0	0.48−1.37
	0		97	0	0.0015	0	0.03−0.46
	10	CuSO₄	20	0.12	0	0	0.45−0.95
	1	FeSO₄	97	−0.105	0	0	0.13−0.22
	0	KMnO₄	25	0.0276	0	0	0.5121
	0	K₂SO₄	35	−0.052	0.16	0	0.13−0.42
	10	MgSO₄	0	0.35	−0.075	0	0.93−2.2
	10		10	0.184	−0.030	0	0.06−2.42
	10		18	0.143	−0.028	0	0.14−2.52
	10		25	0.066	−0.007	0	0.02−2.01
	0		40	−0.024	−0.010	0	0.31−1.56
	0		50	−0.024	−0.006	0	0.32−1.53
	0		60	−0.024	0	0	0.61−1.26
	0		97	0	0.012	0	0.29−0.72
	0		100	−0.030	0	0	1.7294
	10	MnSO₄	25	0.097	−0.017	0	0.24−1.78
	0		35	0.003	−0.016	0	0.43−0.98
	10	NaBrO₃	10	0.09	−0.0085	0	0.4−1.7
	10		25	0	0.0355	0	0.3−0.6
	10	NaCl	15	0.02	0.006	0	1.0−5.6
	10		17.5	0.009	0.007	0	1.6−5.4
	0		38	−0.048	0	0	0.5−5.7
	0		50	−0.048	0	0	1.9−5.9
	0		75	−0.058	0.0055	0	1.8−6.2
	0		94.5	−0.061	0.004	0	3.6−7.9
	0		100	−0.067	0.0051	0	1.8−6.4
	10	NaClO₃	15	0.038	0.0015	0	2.5−7.6
	10		18	0.002	0.014	−0.001	1.1−8.0
	10		25	−0.002	0.0075	0	0.8−4.6
	10	NaClO₄	25	−0.03	0.0105	0	0.6−4.0
	0		25	−0.067	0.0030	0	4.0−16
	0		60	−0.085	−0.0048	0	2.0−22.2
	10	NaHCO₃	25	0.050	0	0	0.6−0.7
	0		50	0.009	0.0085	0	0.4−0.7
	10	NaI	15	−0.017	0.0088	0	3.4−5.6
	0		15	−0.178	0.0068	0	5.6−11.3
	0		25	−0.062	0.0153	0	0.6−3.5
	0		45	−0.099	0.004	0	1.7−7.5
	10	Na₂MoO₄	25	0.018	0.012	0	0.4−1.4
	0		35	−0.050	0	0	0.4−2.6
	0		100	−0.050	0.005	0	1.6−3.8

contd.

Component 1	H_2O	Component 2	t °C	Q_1	Q_2	Q_3	Data for c_2
Na_2SO_4	10	$NaNO_3$	0	0.127	−0.0064	0	2.0−8.4
	10		20	0.037	0.0015	0	2.1−6.7
	10		25	0.021	0.004	0	2.1−4.8
	0		25	−0.113	0.007	0	4.8−5.3
	0		35	−0.022	0.0015	0	2.4−7.0
	0		50	−0.099	0.0070	0	2.3−10.0
	0		75	−0.038	0.0025	0	2.4−16.4
	0		100	−0.038	0.0028	0	2.3−20.5
	10	NaOH	0	0.23	−0.03	0	0.3−7.6
	0		0	−0.046	0	0	7.6−23.4
	10		10	0.041	0.01	0	0.3−4.4
	0		10	−0.044	0	0	6.1−24.4
	10		25	−0.0065	0.031	−0.005	0.9−3.1
	0		25	−0.040	0	0	3.1−26.3
	10		30	0.013	0.024	0	0.5−1.1
	0		30	−0.038	0	0	1.1−20.6
	10	Na_2S	18	−0.11	0.075	0	0.7−2.3
	0		25	−0.031	−0.048	0.031	1.7−2.7
	10	$NaVO_3$	25	0	0	0	0.03−0.15
	10	Tl_2SO_4	25	0.245	0	0	0−0.25
	0		45	0.079	−0.069	0	0.2−0.3
$Na_2S_2O_3$	5	CaS_2O_3	9	−0.037	0	0	0.82−1.76
	5		25	−0.009	0	0	1.42−2.69
	5	NaCl	25	−0.007	0	0	1.0−3.0
	5	$NaNO_3$	25	0.006	0.0015	0	0.9−4.85
$Na_2S_2O_5$	0	Na_2SO_3	25	−0.004	0.004	0	0.2−1.5
	0		35	−0.016	0.027	0	0.2−0.8
$Na_2S_2O_6$	2	BaS_2O_6	0	0.50	0	0	0.09−0.22
	2		12	0.30	−0.31	0	0.30−0.45
	2		20	0.20	−0.20	0	0.15−0.40
$Na_2S_3O_6$	0	$Na_2S_4O_6$	0	0.125	0	0	0.02−1.0
	0		20	−0.087	0.111	−0.031	0.04−1.7
$Na_2S_4O_6$	0	$Na_2S_3O_6$	0	0.022	0.0025	0	1.0−3.0
	0		20	−0.042	0.011	0	0−39
	0	$Na_2S_5O_6$	0	−0.163	0.0457	0.027	0.7−1.13
	0		20	−0.219	0.08	0	0.42−1.5
Na_2SeO_4	0	$CaSeO_4$	25	0.0464	0	0	0.1905
	0	K_2SeO_4	25	0.080	−0.0096	0	1.56−2.1
	10	$MgSeO_4$	25	0.020	0	0	0.83−1.09
	10	NaOH	18	0.020	0.006	0	0.14−4.6
	10		25	0.018	0.012	0	0.17−2.24

contd.

Component 1	H$_2$O	Component 2	t °C	Q_1	Q_2	Q_3	Data for c_2
Na$_2$SeO$_4$	0	NaOH	18	−0.018	−0.0007	0	4.5−26.3
	0		25	−0.0022	0	0	2.24−8.04
NaVO$_3$	0	NaCl	12.5	−0.43	0.055	0	0.4−6.1
	0		60	−0.11	0	0	1.4−6.2
	2	Na$_2$SO$_4$	25	−0.149	0.030	0	0.3−1.9
NiCl$_2$	6	CdCl$_2$	25	0.05	0	0	0.41−0.99
	6	HCl	20	−0.024	0	0	2.6−10.26
	4		20	−0.040	0.0008	0	10.26−15.11
	6		25	−0.0160	0	0	1.8−8.65
	4		25	−0.023	0	0	8.89−13.02
	4		80	−0.023	0	0	0.49−11.02
	6	LiCl	0	−0.076	0	0	3.8175
	4		0	−0.105	0	0	5.4−6.1
	6		25	−0.056	0	0	0.9−1.94
	4		25	−0.070	0.0010	0	3.3−5.7
	2		25	−0.071	0	0	6.94−7.48
	6		17.5	−0.056	0	0	2.6−3.53
	4		17.5	−0.070	0	0	3.18−7.14
	4		50	−0.052	0	0	0.47−4.13
	2		50	−0.064	0	0	2.5−8.13
	2		90	−0.049	0	0	1.98−8.64
Ni(NO$_3$)$_2$	6	HNO$_3$	25	0.006	0	0	2.46−25.78
	4		25	0.0043	0	0	25.78−90.5
NiSO$_4$	7	H$_2$SO$_4$	12.5	0.017	0	0	1.47−2.71
	7		20	0	0.005	0	0.6−1.5
	6		20	−0.013	0	0	1.5−5.07
	7		25	0	0	0	0.74−1.09
	6		25	0.005	0	0	1.09−8.73
	7	K$_2$SO$_4$	25	0.14	0	0	0.32−0.74
NiSeO$_4$	6	H$_2$SeO$_4$	30	0.036	−0.003	0	0.01−5.59
PbBr$_2$	0	LiBr	25	1.15	−0.5	0	0.01−0.82
	0	NaBr	25	1.08	−0.21	−0.07	0.6−1.24
PbCl$_2$	0	LiCl	25	0.025	0	0	0.9−1.68
Pb(NO$_3$)$_2$	0	CsNO$_3$	25	0.30	−0.11	0	0.18−1.46
	0	KNO$_3$	25	0.125	−0.0070	0	0.45−1.6
	0		50	0.082	−0.0035	0	1.17−5.28
	0		100	0.035	0.0005	0	0.87−4.63
	0	LiNO$_3$	25	−0.03	0.0030	0	0.12−10.3
	0	NH$_4$NO$_3$	10	0.115	−0.005	0	1.99−21.52
	0		20	0.07	−0.0018	0	2.2−10.0
	0		25	0.07	−0.0017	0	2.8

contd.

Component 1	H_2O	Component 2	$t\,C°$	Q_1	Q_2	Q_3	Data for c_2
$Pb(NO_3)_2$	0 0	$NaNO_3$	0 50	0.022 0.004	0.0033 0.002	0 0	0.4802 0.91 — 7.43
RbBr	0	$MgBr_2$	25	−0.054	−0.004	0	1.13 — 4.37
RbCl	0	$BeCl_2$	25	−0.032	0	0	0.57 — 8.4
	0	$CoCl_2$	25	0.014	0	0	1.11 — 2.39
	0 0	LiCl	25 40	−0.033 −0.065	0 0	0 0	0.56 — 1.47 0.55 — 2.04
$SrBr_2$	6	HBr	25	−0.018	−0.0008	0	0.91 — 3.99
$SrCl_2$	6 2	$CoCl_2$	25 50	0.005 0.030	0 −0.0025	0 0	1.76 — 4.62 3.95 — 7.31
	6	HCl	25	−0.030	0	0	0.27 — 10.51
	6 6 2 2	KCl	18 60 60 100	0.040 0.040 0.040 0.040	−0.0007 0.0040 −0.0045 −0.0045	0 0 0 0	0.97 — 1.56 0.67 — 0.70 0.70 — 1.65 0.7 — 1.95
	6 6 2	LiCl	25 40 40	−0.112 −0.083 −0.35	0.0150 0.014 0.034	0 0 0	0.94 — 4.44 0.45 — 3.25 3.25 — 8.0
	6 6 2 2	NaCl	18 60 60 100	0.0044 0.0415 0.120 0.013	0 0 −0.092 0	0 0 0 0	1.4 — 2.3 0.16 0.3 — 0.9 1.25 — 1.27
	6	$Sr(NO_3)_2$	25	0.046	−0.006	0	1.39 — 2.90
$Sr(NO_3)_2$	4 0	$Ca(NO_3)_2$	60 60	−0.013 −0.013	−0.002 −0.001	0 0	1.3 — 8.07 3.9 — 20.4
	4 4 4 0 4	HNO_3	0 15 25 25 50	−0.111 −0.055 −0.055 0.17 −0.061	0.0100 0.0200 0.0015 −0.043 0.0015	0 0 0 0 0	0.68 — 9.58 1.93 — 6.58 0.52 — 3.00 3.0 — 10.91 1.4 — 27.5
	4 4	KNO_3	20 40	0.070 0.045	−0.0025 −0.0017	0 0	2.0 — 4.77 2.48 — 7.69
	4	$SrCl_2$	25	−0.050	0	0	0.88 — 1.25
SrS_2O_6	4	$(NH_4)_2S_2O_6$	30	0.14	−0.015	0	1.15 — 8.31
TlCN	0 0	KCN	25 25	0 0.0366	0.0045 0	0 0	0.06 — 6.39 7.16 — 9.37
$TlNO_2$	0	$Ba(NO_2)_2$	25	0.185	−0.01	0	1.28 — 2.81

cond.

Component 1	H$_2$O	Component 2	t °C	Q_1	Q_2	Q_3	Data for c_2
TlNO$_2$	0	KNO$_2$	25	0.062	−0.0007	0	4.8−9.1
	0	NaNO$_2$	25	0.043	0	0	4.0−13.2
Tl$_2$SO$_4$	0	Na$_2$SO$_4$	25	0.895	−0.5	0.205	0.45−1.75
	0		45	0.790	−0.5	0.16	0.7−3.36
UO$_2$(NO$_3$)$_2$	6	KNO$_3$	0	0.070	−0.038	0	0.5428
	6		5	0.070	−0.0064	0	1.0456
	6		10	0.070	0.0027	0	1.3962
	6		15	0.070	0.0022	0	1.8186
	6		20	0.070	−0.0009	0	2.3472
	6		25	0.057	0	0	0.79−3.08
	6		50	0.017	0	0	0.6−8.94
	6		70	0.017	0	0	0.9−12.98
	6		90	0.017	0	0	0.8−18.1
	6	NH$_4$NO$_3$	25	0.040	0	0	0.7−10.63
	6	NaNO$_3$	25	0.025	0	0	0.9−4.8
UO$_2$SO$_4$	3	K$_2$SO$_4$	25	0.161	0	0	0.0343
ZnCl$_2$	0	BeCl$_2$	25	−0.07	0.002	0	2.3−8.49
	0	CaCl$_2$	25	0.0022	0	0	11.71
	0		60	−0.0160	0	0	0.4−11.8
	1.5	LiCl	25	−0.026	0	0	2.08−5.54
	0		40	−0.011	0	0	2.5−11.05
	0	TlCl	25	−0.017	−0.006	0	0.9−2.05
Zn(NO$_3$)$_2$	6	HNO$_3$	20	0.015	0	0	3.97−18.56
ZnSO$_4$	7	BeSO$_4$	0	−0.038	0	0	0.21−2.74
	7		25	−0.0235	0	0	0.24−2.52
	6		50	−0.014	0	0	0.28−0.73
	1		50	−0.028	0	0	0.73−2.76
	1		99	−0.028	0	0	0.3−6.92
	7	H$_2$SO$_4$	−4.5	−0.035	0.005	0	0.02−7.48
	7		6	−0.020	0.005	0	0.7−6.53
	7		15	−0.012	0.006	0	0.015−5.24
	6		15	0	0	0	5.24−8.95
	7		25	0.007	0.005	0	0.01−3.3
	6		25	0.006	0	0	3.3−3.78
	7		35	0.007	0	0	0.03−0.9
	6		35	0.006	0	0	0.9−2.59
	6		45	0.030	0	0	0.07−1.14
	1		50	−0.02	0	0	0.14−21.8
	1		65	−0.01	−0.0015	0	0.07−15.25
	1		70	0	−0.0027	0	0.05−12.7
	1	K$_2$SO$_4$	80	0.085	−0.036	0	0.35−0.68
	1		100	0.092	0	0	0.21−1.55

Subject index

activity 23, 25, 144, 150, 157, 177
adhering mother liquor 88
adsorption inclusion 178
analytical methods 85
anhydrous projection 41, 47, 83
anomalous mixed crystals 178
aqueous projection 41, 47, 83

basic equation of phase equilibria 21
binary systems 32, 60, 181

central projection 111, 122, 127
chemical potential 20, 23, 150, 157, 158, 177
Clapeyron equation 18
Clausius-Clapeyron equation 23
clinogonial projection 132
component 27, 29
concentration 30, 93
condensed system 27
conductometric measurement 94, 99
conformal solutions 147
congruent point 64
conjugated solutions 62, 70
conode 77
cooling curve 100
correction of cooling curves 102
critical dissolution temperature 62

Debye-Hückel equation 141
degree of freedom 27
Doerner-Hoskins distribution law 158, 177
dynamic methods 99

enantiotropy 58
enthalpy diagram 50
entropy of fusion 18
establishment of equilibrium 86
eutectic point 61, 73
eutectic temperature 61
eutonic point 41, 86

four-component system 44, 79, 116

Gibbs-Duhem equation 20, 24
Gibbs phase rule 26, 32, 33, 38
Gromakov method 107
Guldberg-Waage law 176

Hahn precipitation rule 176
heat of dissolution 22, 25, 26
heat of fusion 18, 22
homogeneous distribution law 178
hydrates 74, 76, 80, 89

ideal solubility 22
inclusion 176, 179
incongruent point 65
incorporation of impurities 176
interaction constant 162
internal adsorption 179
invariant point 39, 83
isohydric solutions 155
isomorphous incorporation 176
isopiestic solutions 152, 155
isothermal methods 98

Jänecke's projection 42, 45

kinetics of dissolution 103
kinetics of growth 103

le Chatelier principle 22
lever rule 49
liquidus 66
logarithmic distribution law 177
lower solubility limit 179

macro-component 177, 178
material balance 20
mechanical inclusions 179
melt 29, 58

melting point 22, 61, 64, 100
metastable modification 59
micro-component 177, 178
mixed crystals 25
modification change 19, 23, 58
molality 31, 161
monotropy 60
multicomponent systems 19, 48, 119, 162

Nernst distribution law 177

orthogonal projection 107
Ostwald rule 59
Othmers correlation method 22

Paneth rule 178
partial molar quantities 19
peritectic point 66
peritectic temperature 66
phase 27
polythermal method 92
pressure 17, 18, 19, 21, 56

reciprocal salt pair system 28, 45, 82, 148
refractive index 94
relative activity coefficient 158
relative molality 158
Richards rule 18

sampling 87
Schreinemaker's method 91
sectorial growth 179
selective adsorption 179

single-component systems 17, 32, 56
solid solution 29, 43, 70, 102, 159
solidus 66
solubility product 160
solution 29
straight line rule 49, 91
sublimation 19, 57
supercooling 95
supersaturation 95, 104
symmetrical correlation method 135
synthetic methods 92

temperature 17, 19, 21, 24, 26, 33, 39, 47, 56, 72, 85, 93, 113, 175
ternary diagram 34, 35, 36, 37, 38
theory of electrolyte solutions 141
thermometric method 99
three-component system 33, 72, 91, 107, 111, 205
Töppler method 95
transformation of coordinates 42
transition point 19
triple point 57

van't Hoff coordinates 42

water of hydration 23
weighing of components 86

x—y diagram 78

Zdanovskii method 150

/541.34N999S>C1/

DATE DUE

DEC 11 1997

Demco, Inc. 38-293